Structured Systems Analysis
and Design Methodology

INFORMATION SYSTEMS SERIES

Consulting Editors

D.E. AVISON BA, MSc, FBCS
*Department of Computer Science and
Applied Mathematics, Aston University
Birmingham, UK*

G. FITZGERALD BA, MSc, MBCS
*Oxford Institute of Information Management
Templeton College, Oxford, UK*

This series of student texts covers a wide variety
of topics relating to information systems. It is
designed to fulfil the needs of the growing
number of courses on, and interest in, computing
and information systems which do not focus
purely on the technological aspects, but seek to
relate these to business or organisational context.

Structured Systems Analysis and Design Methodology

GEOFF CUTTS BSc, MSc, CEng, FBCS
School of Computing and Management Sciences
Sheffield City Polytechnic

SECOND EDITION

OXFORD
BLACKWELL SCIENTIFIC PUBLICATIONS
LONDON EDINBURGH BOSTON
MELBOURNE PARIS BERLIN VIENNA

Blackwell Scientific Publications
Editorial offices:
Osney Mead, Oxford OX2 0EL
25 John Street, London WC1N 2BL
23 Ainslie Place, Edinburgh EH3 6AJ
3 Cambridge Center, Cambridge
   Massachusetts 02142, USA
54 University Street, Carlton
   Victoria 3053, Australia

Other editorial offices:
Arnette SA
2, rue Casimir-Delavigne
75006 Paris
France

Blackwell Wissenschaft
Meinekestrasse 4
D-1000 Berlin 15
Germany

Blackwell MZV
Feldgasse 13
A-1238 Wien
Austria

First edition published by
   Paradigm Publishing Ltd   1987
Reprinted 1988
Reprinted by Blackwell Scientific
   Publications 1989
Second edition published 1991

Set by Excel Typesetters Co.
Printed and bound in Great Britain by
Billings & Son, Worcester

DISTRIBUTORS
   Marston Book Services Ltd
   PO Box 87
   Oxford OX2 0DT
   (*Orders*: Tel: 0865 791155
          Fax: 0865 791927
          Telex: 837515)

USA
   Blackwell Scientific Publications, Inc.
   3 Cambridge Center
   Cambridge, MA 02142
   (*Orders*: Tel: (800) 759-6102)

Canada
   Oxford University Press
   70 Wynford Drive
   Don Mills
   Ontario M3C 1J9
   (*Orders*: Tel: (416) 441-2941)

Australia
   Blackwell Scientific Publications
   (Australia) Pty Ltd
   54 University Street
   Carlton, Victoria 3053
   (*Orders*: Tel: (03) 347-0300)

British Library
Cataloguing in Publication Data
Cutts, Geoff
   Structured systems analysis and design
   methodology.—2nd. ed.
   1. Computer systems. Structured
   systems analysis
   I. Title
   004.21

   ISBN 0-632-02831-9

Library of Congress
Cataloging in Publication Data
Cutts, Geoff.
      Structured systems analysis and design
   methodology/Geoff Cutts.—2nd ed.
      p. cm.
      Includes index.
      ISBN 0-632-02831-9
      1. Electronic data processing—
   Structured techniques.   I. Title.
   QA76.9.S84C88   1991
   005.1'13—dc20

Contents

Contents

ix

Foreword

The Blackwell Scientific Publications Series on Information Systems is a series of student texts covering a wide variety of topics relating to information systems. It is designed to fulfil the needs of the growing number of courses on, and interest in, computing and information systems which do not focus on the purely technological aspects, but seek to relate these to the business and organisational context.

Information systems has been defined as the effective design, delivery, use and impact of information technology in organisations and society. Utilising this fairly wide definition, it is clear that the subject area is somewhat interdisciplinary. Thus the series seeks to integrate technological disciplines with management and other disciplines, for example, psychology and philosophy. It is felt that these areas do not have a natural home, they are rarely represented by single departments in polytechnics and universities, and to put such books into a purely computer science or management series restricts potential readership and the benefits that such texts can provide. This series on information systems now provides such a home.

The books will be mainly student texts, although certain topics may be dealt with at a deeper, more research-oriented level.

The series is expected to include the following areas, although this is not an exhaustive list: information systems development methodologies, office information systems, management information systems, decision support systems, information modelling and databases, systems theory, human aspects and the human-computer interface, application systems, technology strategy, planning and control, and expert systems, know-ledge acquisition and representation.

A mention of the books so far published in the series gives a 'flavour' of the richness of the information systems world. *Information Systems*

Development: Methodologies, Techniques and Tools (D.E. Avison and G. Fitzgerald), looks at many of the areas discussed above in overview form; *Information and Data Modelling* (David Benyon), concerns itself with one very important aspect, the world of data, in some depth; *Multiview* (Avison and Wood-Harper) describes a contingent approach to information system development; *Information Systems Research: Issues, Techniques and Practical Guidelines* (R. Galliers, (ed)) provides a collection of papers on key information systems issues; and *Software Engineering for Information Systems* (D. McDermid) discusses software engineering in the context of information systems. There are a number of other titles in preparation.

This latest addition to the series is a new edition of Geoff Cutts' successful book *Structured Systems Analysis and Design Methodology*. The book is clearly in the mainstream of information systems and is a very welcome addition to the series. Geoff Cutts describes a 'generic' structured systems analysis and design methodology. There are many different versions of structured systems methodologies and readers will meet a wide variety of them in industry and education. The philosophy of this book is to describe and demonstrate the common or generic features and not describe a particular commercial methodology. This enables Geoff to clearly focus on the rationale, benefits, methods and techniques of a structured or disciplined approach to information systems development.

It is felt that this is the correct approach for both students and practitioners, enabling them to understand the principles (which are not always explicitly discussed in commercial methodologies) and then be in a position to place in context any commercial structured systems product that may be encountered. Further, many organisations evolve their own methodologies based on a variety of products, adopting, changing and simplifying as appropriate to their particular requirements and environment.

The philosophy of this book should make an informed contribution to that process. Although the book is not a description of any commercial product it references particular products such as Systems Engineering and SSADM Version 4 where appropriate. Thus the book is a timely and important contribution to the literature on methodologies and in particular provides a simplified approach to the structured paradigm which makes it easy to follow, teach and learn. It has been observed that the take-up of such methodologies has been relatively slow and it is hoped that this text will help reduce the complexity and help separate the methodological wood from the trees.

David Avison and Guy Fitzgerald
Consulting Editors
Information Systems Series

Preface

The purpose of this book is to explain a structured systems analysis and design methodology in a highly pragmatic fashion. I have tried to use my personal experiences as an analyst, designer and manager of many business development projects, together with my academic studies, to produce a book which would be helpful to potential computer users, existing computer users, systems managers, systems analysts, systems designers and students of analysis and design. I have received many encouraging comments on the first edition of the book. In this second edition I have tried to incorporate material based on these comments.

The book comprises three parts. Part 1 introduces structured systems analysis and design. It describes the need for a different approach by the computing industry, the stages, parts, tasks and techniques and the benefits of adopting such an approach. All classes of reader should appreciate the need for a different approach to systems analysis and design. Part 1 assesses this issue and should be studied by all readers.

Part 2 describes the three stages and could be used for education and for reference by the practising analyst and designer, as well as the student of analysis and design. It describes in detail the stages, parts and tasks which comprise the methodology. Systems analysts, systems designers and students of systems analysis and design should study this part in detail. User and management involvement in computer development projects is a vital ingredient for success. Users and management should understand the tasks and techniques to be able to contribute fully. Part 2 should be read by users and management to gain this understanding.

No methodology can be mastered without practice. Part 3 provides a detailed case study which may be used to practice the methodology.

The methodology described is derived from LSDM: Learmonth and Burchett Structured Development Method. It is a structured approach to systems analysis and design that will be of use to everyone interested in a disciplined approach to systems development.

The benefits accrued from adopting a disciplined approach, particularly a structured approach, are explained in Chapter 1. The review of the first edition in *Computing* concluded by saying 'Geoff Cutts does a good, well-organised job in explaining this difficult material and as such I would recommend the book as suitable reading for anyone interested in a disciplined approach to systems development'.

All organisations should be interested in improving quality and productivity. This text is a first introduction to a disciplined approach to systems development and forms a basis for adoption of a structured methodology, whether the method be that described in the text or one of the many commercially available methods such as Learmonth and Burchett Structured Development Methods (LSDM) and Systems Engineering and the CCTA's SSADM.

The Guardian said of the first edition 'The book is well written and well constructed; where the text needs a diagram to illustrate a point the diagram usually exists and actually does illustrate the point in question. It would be a good book for a college course on systems analysis and design'.

The book is aimed at organisations wishing to look towards a structured approach and at University and Polytechnic courses at undergraduate and post graduate level. It provides the main input to a course on structured systems analysis and design introducing a disciplined approach via the introduction of a methodology. Students should not be introduced to a single methodology; this text will enable them to understand the structured approach, to practise the approach and to master the techniques used in the methodology. It will further enable the student to examine commercially available methodologies.

The text describes a structured methodology, one of the themes that has evolved in Information Systems Methodologies. Other themes have developed; for an overview of methodologies and a look at the themes the reader is recommended to read *Information Systems Development: Methodologies, Techniques and Tools* by D.E. Avison and G. Fitzgerald.

This second edition of my book brings the stages and parts in the methodology more into line with current practice: Stage A, Analysis; Stage B, Logical Design; and Stage C, Physical Design. The consideration of business options is placed earlier in the methodology giving increased emphasis to user and senior management involvement and consequently contribution to the very important decision making process. The contract between the user and developer, describing the requirements exactly, is the vital document.

In detail this second edition structures the tasks into three stages and five parts, placing more emphasis on user and management involvement with the development process. It introduces a data dictionary and dialogue specification into a detailed business specification. Prototyping is discussed in the way that it contributes to the structured approach.

The major development is within the specification of requirements part of the methodology. The emphasis is placed on the creation of an outline business specification involving the postulation and review of business options before the design constraints are set by management in terms of time, money and requirements, and a detailed business specification is produced. The business specification is then submitted to the design and implementation phases.

Geoff Cutts

Acknowledgements

First Edition

This book is a result of many years' experience with systems analysis and design. My experience with International Computers Limited and as Systems Development Manager for GMS Computing Ltd. provides the industrial input to the book. My recent research, consultancy and study of systems analysis and design provide the academic input to the book.

The support of Learmonth and Burchett Management Systems made the book possible. The synopsis for the book was developed in conjunction with LBMS who also provided assistance by reviewing the book, by allowing me to attend some of their courses and by allowing their package Auto-Mate to be used for the development of the case study.

I have found that writing a book of this size is a large task. I must acknowledge the love and support of my family, Janet, Gareth and Nicky, and the tremendous efforts of Louise Morrey who tirelessly word processed my manuscript to produce the book. A sincere thank you to my wife and my father who both gave many hours to the task of proof reading.

Second Edition

The second edition incorporates many suggestions and improvements from staff and students at Sheffield City Polytechnic and from delegates on short courses. I have also tried to incorporate some of the experience of consultancy projects since the first edition. I am particularly grateful to Sue Morton for all her helpful suggestions and to LBMS for the information on Systems Engineering. May I also thank David Avison and Carol Byde who read and commented on the text during its production.

I must thank my wife and father who again gave their time for proof reading, and my wife and Joanne Willison who diligently word processed the second edition.

PART 1

BACKGROUND AND OVERVIEW

Chapter 1
The Environment of Analysis and Design

1.1 INTRODUCTION

The development of a large computer system is one of the most complex activities undertaken by organisations. The number of staff involved and the resources consumed often make computer development projects one of the most costly of all projects undertaken.

Projects which exceed their projected development costs and their projected development timescales are very often the norm. In addition, many of these systems do not provide the facilities the user required.

Improvements in the productivity of staff involved with systems development and the production of better quality systems are vital if systems engineering is to be accepted as a true engineering discipline.

The emergence of methodologies in the late 1960s and 1970s began the trend away from the notion that good programming was all that was important. There was a realisation that good analysis and design was equally, if not more, important.

Techniques for programming, such as the many versions of structured programming, are now well established in organisations. Improvements in productivity and quality cannot be achieved by the introduction of structured programming if the design specification from which the system is programmed and implemented is incomplete, ambiguous and contradictory.

Many of the problems stem from the use of natural language as the only language for design specifications. Many problems result from the lack of a methodology for analysis and design leading to the production of the design specification.

The 1980s have seen the growth of a large number and variety of information systems methodologies. The methodologies fit into a number of important themes.

The systems approach concerns itself with interaction between the system and its environment; information systems concern people as well

3

as technology. Whatever methodology is adopted the systems analyst should be aware of the systems approach. Checkland (1981) developed a practical methodology, Soft Systems Methodology, SSM, based on the systems approach.

The participative approach involves all those people affected by the new system, encouraging user involvement. Effective Technical and Human Implementation of Computer-Based Systems, ETHICS, developed by Mumford (1979), is a methodology based upon the participative approach.

Structured methodologies are based on decomposition, breaking down a complex system in a disciplined way into manageable units. Structured systems analysis and design has been developed by a number of consultants and authors. It emerged with the work of Gane and Sarson (1979) and De Marco (1979). These texts were in addition to the work of Yourdon and Constantine (1978), Myers (1978) and Jackson (1983).

The techniques of systems analysis and design such as data flow diagrams, entity models and normalisation appear in many of the methodologies. The techniques may be used in different ways with different interpretations within each methodology. Broadly, each technique has a common principle that is important.

In 1980 Learmonth and Burchett Management Systems (LBMS) were invited with the Central Computer and Telecommunications Agency (CCTA) to carry out a joint project to develop a standard analysis and design method for central government. This resulted in Structured Systems Analysis and Design Method, SSADM, which is now the standard for government projects. Version 3 of SSADM is described in Ashworth and Goodland (1990). LBMS market the method as LSDM, LBMS Structured Development Method.

There were initially a number of minor differences between SSADM and LSDM. Both have been subject to continual review and development with the result that in 1990 there exists a version 4 of SSADM and LBMS Systems Engineering.

The array of commercially available methodologies is ever increasing with many methodologies and multiple versions of each. The current problems with systems engineering provided the motivation for these developments. This book describes a methodology for systems analysis and design leading to a design specification which is specific, complete, unambiguous, non-contradictory, clear and concise. The book attempts to give the reader an understanding of the structured approach through the fitting of standard techniques into a structured methodology.

The use of techniques such as data flow diagrams and entity modelling for systems engineering is well established. Many organisations use their own methods. What is not so common is the structuring of the techniques into a practical methodology for system engineering. Structured systems analysis and design provides a methodology, incorporating the best available techniques, for the analysis and design of systems. Structured systems analysis and design is not a collection of techniques, it is a well-developed step by step approach, which commences with an investigation of the current system and concludes with a detailed system design specification.

The methodology does not concentrate on any one technique. Some methodologies concentrate on functional analysis using data flow diagrams; some concentrate on data analysis using data models. Structured systems analysis and design regards functions and data with equal importance and uses a variety of techniques, some within-more than one stage, to support the objectives of the stage. The techniques chosen for each task are the latest techniques which support the achievement of the task's objectives. Structured systems analysis and design is a structuring of well-known techniques into a comprehensive methodology which solves many of the problems associated with systems engineering.

1.2 THE PROBLEMS ASSOCIATED WITH SYSTEMS ENGINEERING

The major problems associated with systems engineering can be deduced from the symptoms evident in many development projects. These symptoms are:
• development costs over budget
• development timescales over planned timescales
• the production of systems which do not meet the user's requirements and which are difficult to modify and maintain.

The massive increase in the size and complexity of systems being developed without any comparable change in the methods for analysis and design is perhaps the major reason for the problems being experienced by many organisations.

Point of agreement	Increase in cost
At the specification of requirements phase	None
At the end of the design phase	The cost to rework the analysis and design
At the end of programming	The cost to rework the analysis, design and programming
During user testing	The cost to rework the analysis, design, programming and testing
During live running by maintenance and modification	The very considerable costs required to rework the analysis and design and to modify the existing programs

Fig. 1.1

The system implemented must meet the user's requirements. The requirements must also be met within the cost and timescales agreed for the system development process. The cost and time of meeting the requirements increases in proportion to the time taken to reach the point where a specific, complete, unambiguous and clear statement of them is agreed with the user. Figure 1.1 shows this increase in cost.

It is vital to remove the errors from the design at an early stage. It is also important to accept that users' requirements will change as their environment changes. Maintenance and modification to all systems will be necessary. The methodology, therefore, must facilitate improvements in productivity and in the quality of development projects,

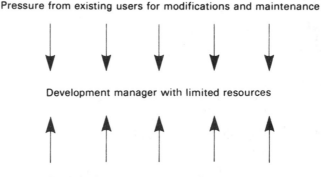

Pressure from existing users for modifications and maintenance

Development manager with limited resources

Pressure from potential users for new systems

Fig. 1.2

maintenance projects and modification projects. Figure 1.2 shows the current position for many development managers.

The problem of proving the correctness of analysis and design is made difficult by the lack of a notation common to all stages of development. Changes from one notation to another as the development process progresses from stage to stage introduce errors of interpretation. If the statement of requirements is written in a vague fashion, using natural language only, it is impossible to ensure the design meets the statement of requirements. Figure 1.3 shows typical proportions of the results of developing new systems.

A specification of requirements written in natural language is shown below. It is vague, contradictory and incomplete.

Specification of requirements
(1) The new system must provide a fully on-line service for all sales accounting functions.
(2) Statements should be printed at regular intervals showing all invoices and payments.

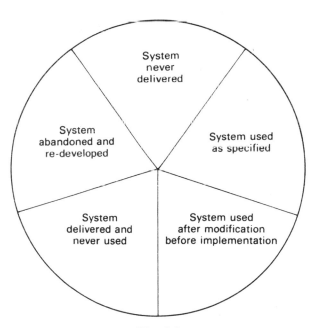

Fig. 1.3

(3) Each delivery should result in an invoice.
(4) Invoices should be printed weekly showing all deliveries made to a customer during the week.
(5) Credit notes should be treated as negative invoices.

Vague
- define regular interval – requirement 2
- are invoices and payments printed forever? – requirement 2

Contradictory
- each delivery = invoice – requirement 3 *or*
  invoice = all deliveries for one week – requirement 4
- are credit notes printed? – requirements 2 and 5

Incomplete
- how are new customers inserted into the system?
- how are invoices priced?

A methodology for the development of the specification of requirements and the design specification is required to overcome these problems.

A methodology is simply a way of structuring one's thinking about an area of study. It ensures that the appropriate sub-areas are considered at the appropriate time, together with the proper reasoning about them. Further, a methodology should provide a framework which leads towards the achievement of specific objectives. The achievement of the objectives is not guaranteed; it depends on a number of criteria.

Questions such as 'Under what conditions is it sensible to seek to achieve the objectives?' and 'Do these conditions apply to the area of concern?' need to be asked and positively answered. Finally, one should ask if the skills and resources are available to implement the methodology. These questions and others are considered now.

1.3 THE PROBLEMS ASSOCIATED WITH SYSTEMS ANALYSIS

The major problems associated with systems analysis arise in the understanding of the user's environment and the subsequent specification of the user's requirements. Both of these problem areas arise from difficulties in communication between the user and the systems analyst.

The problem area is often considerably exasperated by the continuous change of the user's environment and the user's requirements for the new system.

The systems analyst can find it very hard to learn sufficient about the business to observe the system from the user's viewpoint. Similarly, the user community does not know enough about computer systems to be able to specify the requirements accurately, unambiguously and precisely. The analyst has no right to expect a lucid explanation of the systems requirements. The user has no right to expect a system which meets the requirements without such an explanation and subsequent documentation.

The analysis task is therefore a partnership between user and analyst which the analyst is usually asked to manage.

A large part of the investigation and analysis phases within any project is spent acquiring detailed information about the current system. During these phases the analyst can quickly get overwhelmed with detail. Unless there is one scheme or structure to organise this detail, the analyst can become overloaded with facts and paper. The detail is needed and must be available when required, but the analyst must have techniques and tools to control it.

The document setting out the detail of the new system effectively forms a contract between the user and the project team. If the document is impossible for the user to understand because of its sheer size and technical content, then no contract exists and the user possesses a free hand to change the requirements. Changing requirements during the design and implementation phases is one of the major reasons for project delay and overspend.

1.4 THE PROBLEMS ASSOCIATED WITH SYSTEMS DESIGN

Problems associated with the user's understanding of the specification documents were discussed in the previous section. One solution might be to write a business specification using the user's terminology, which would be totally understandable to the user. However, if a specification document were written in such a way it might not be very useful to the designers and programmers who will be involved during the later phases of the project.

Often a considerable amount of re-analysis takes place defining data and functions in design terms. This work essentially duplicates the work of the analyst.

The second major problem for the designer is the separation of system design, data and processes from technical design. Exactly how can the design be implemented on the given hardware and software? The designer needs to be able to design a logical database and logical processes before the technical detail is added to the documentation.

What is required, therefore, is a systems analysis and design methodology which, as far as possible, overcomes these problems.

1.5 THE NEED FOR A STRUCTURED METHODOLOGY

The need for a structured methodology is evident from the many sets of survey results which report the ever increasing problems with current methods of developing computer systems. The major problem is related to project costs and timescales. Only a small proportion of projects are completed within their budget and on time. There are many reasons given for this situation, ranging from poor estimating to continued change in the user's requirements. In addition to the problems of cost and timescale, systems are still developed that do not meet the user's requirements, are difficult and costly to modify and maintain and are poorly documented.

The need for a structured methodology is not solely based on the problems associated with current approaches. There exists a requirement in all areas of computing to improve the productivity of development staff and the quality of the final product. Productivity and quality may be partially linked to user and management involvement; they should not, however, depend entirely on the involvement of users and management. A requirement exists in parallel with the productivity and quality requirement, to provide a basis for progress measurement and resource planning.

Published statistics by many organisations show that very few systems, as initially implemented, meet the user's requirements. The cost of rectification increases from zero, if the user's requirements are specified exactly during the analysis phase of a project, to very large sums if the system is implemented and then modified to meet the requirements. The reduction of this rectification cost is another reason to follow a structured methodology.

1.6 CHARACTERISTICS OF A 'GOOD' SYSTEMS ANALYSIS AND DESIGN METHODOLOGY

This initial chapter has established a need for a structured methodology which will contribute towards improved development productivity and system quality. It is now necessary to establish a set of characteristics of a 'good' systems analysis and design methodology against which structured systems analysis and design might be measured.

Two important characteristics of any systems analysis and design methodology are the degree to which they assist analyst – designer – implementor communication and the ease of understanding both of the methodology itself and of the documentation produced by it. The systems being considered for the 1990s and beyond are large and complex. Therefore, to aid understanding, some form of top – down modelling is required. The modelling techniques must provide high level views which can be exploded piece by piece to provide more and more detailed local views. The view of a system must encompass three basic concepts: a view of the data within a system, a view of the flow of information around the constituent parts of the system, and a view representing the effect of information flow or functions on the data due to the passage of time.

The development of these three views of the system can be undertaken in a structured manner if the methodology provides for decomposition of the system. In this way a manageable number of system elements can be considered at one time.

The concept of considering a small number of elements at any time ensures that they are carefully selected for analysis. Elements may comprise functions and data structures. This process effectively defines the boundary of the project, all selected elements being initially inside the boundary.

Understanding also depends on clarity of expression. A good methodology should, wherever possible, provide graphical representations. Such representations are capable of conveying large quantities of information in a simple and concise form. This makes the models represented graphically easy for users to understand and relatively easy to change. The ability to gradually transform a set of diagrams from a representation of the current system through a representation of the requirements towards a detailed design specification is a vital design characteristic. This ensures the methodology leads from analysis of the current system into requirements specification and on into detailed design specification.

1.7 THE SPECIFICATION

Two formally documented specifications are necessary: the specification of requirements and the detailed design specification. In each case a good methodology should lead to concise, unambiguous specifications which are not excessively wordy, physical or redundant. Many specifications are unclear, inconsistent and incorrect. The product of a structured methodology should be graphic and concise and overall it should accurately reflect the user's requirements. The specifications must also be totally understandable by their recipients.

1.8 THE SYSTEMS DEVELOPMENT CYCLE

There are many different representations of the systems development cycle. The following phases are used in this book:
(1) terms of reference, including business objectives
(2) feasibility study
(3) investigation
(4) analysis
(5) design
(6) implementation
(7) operation, maintenance and modification.

All projects must pass through these phases, with a different emphasis being placed on each according to the type of project. Very often the first phase has been completed before the project commences, with the terms of reference and business objectives forming the project brief. The first phase in these cases is a feasibility study. No feasibility study should commence without clearly stated terms of reference and overall business objectives.

1.8.1 Feasibility study

A feasibility study is required to determine if the objectives are realistic, i.e., they are able to be done and put into effect. The feasibility study must address technical, economic and social areas of concern.

At the end of a feasibility study, a stop/go decision must be made for the project. The study must therefore present a detailed specification of the objectives, together with an analysis of the impact on the organisation of any project aimed at meeting them.

Techniques such as cost benefit analysis, discounted cash flow and return on investments are used to establish economic feasibility. Technical feasibility needs to be established by estimation of processing and data requirements, while social feasibility can only be determined by discussions with the proposed user.

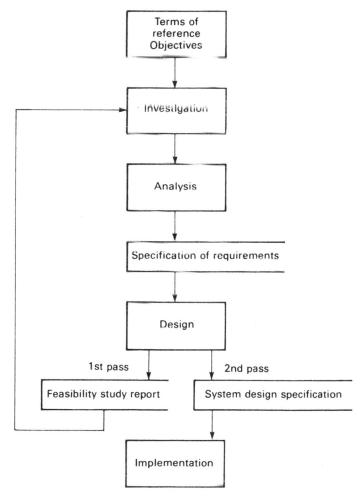

Fig. 1.4 The two-pass systems development cycle

The feasibility study is a project within a project incorporating investigation, analysis, design and possibly a pilot implementation/ operation. A good methodology should provide for high-level investigation, analysis and design leading to the feasibility study report, followed by more detailed investigation, analysis and design leading to the final system design specification. This is termed the two pass systems development cycle and is shown in Figure 1.4.

The feasibility study should refine the business objectives within the terms of reference to provide objective measures against which success or failure of the project may be judged. It should further consider alternative system designs, rejecting unacceptable ones according to economic, technical and social feasibility criteria, finally choosing a number of designs for development. The feasibility study should report when a forecast can be made of the effect on the objective measures previously defined.

At this stage the feasibility study report should lay the foundations for subsequent planning and control of the project by documenting the investigation, analysis and design work done so far. This documentation should allow any subsequent project team to benefit from the work of the feasibility study team.

1.8.2 Investigation

The investigation phase comprises a detailed study of the existing system. Investigation continues until a detailed model of how the current system is implemented can be drawn and agreed with the user. The model should include both the functions and data within the system and the volume of processing and data. Finally, the identified problems with the current system and any requirements for the new system should be documented.

Investigation techniques include the study of existing data, i.e. documents and reports, observation, the use of questionnaires and, of course, interviewing.

1.8.3 Analysis

There are very many different definitions of systems analysis with no real common agreement about the precise meaning of the term. One of the reasons may be the diversity of systems themselves, making a common approach impossible.

One useful definition is 'the organisation of information gathered during the investigation phase into a meaningful form'. This generally means the building of a model which represents what the current system accomplishes, not how it is accomplished. The model is logical not physical, and should include both the functions carried out by the system and the data stored within the system.

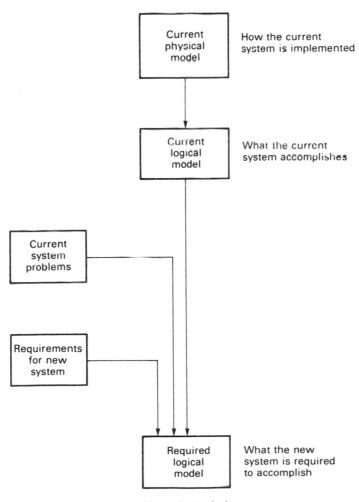

Fig. 1.5 Analysis

The model first reflects the current system. However, the definition may be enhanced to include the improvement and optimisation of the model to solve the identified problems with the current system. Further, the model should be enhanced to provide the stated requirements, making the final product of analysis a required logical model or specification of requirements. This process is shown in Figure 1.5.

1.8.4 Design

The specification of requirements forms the input to the design phase. The objective of this phase is to specify the system in a way suitable for implementation on the chosen hardware with the chosen software. The design phase may include the choice of hardware and software.

The design phase may be divided into logical design and physical design. The logical design sub-phase will transform the specification of requirements into detailed specifications of both data and processing requirements: what, in detail, is required of the new system. The physical design sub-phase will transform the logical specifications into physical specifications, database and program specifications, by including, within the specifications, consideration of the constraints imposed by the chosen hardware and software.

During the physical design sub-phase, details such as input and output formats will be designed along with file formats, screen formats etc.

The design phase will conclude with the production of a detailed system design specification.

1.8.5 Implementation

Implementation is the phase where the system design is translated into an operational system. It includes programming, program testing, hardware and software acquisition and installation, system testing, system set-up, user education and training, user testing and change-over to the new system.

Programming and program testing are sometimes identified as a separate phase. The use of structured systems analysis and design will reduce the reliance on program and system testing to identify analysis and design errors. For this reason, programming and program testing are included as a necessary part of the implementation phase without any special importance.

1.8.6 Operation

The operational life of any system should include its review at agreed intervals. Business objectives will have changed during the life of the system and the degree with which they are met by the operational system should be established. Operational problems and requests should also be considered so that a planned modification and maintenance programme for the system can be established.

All systems require unplanned emergency maintenance. This area of work for analysts, designers and programmers is consistently increasing. The structured approach aims to reduce this effort by ensuring high quality, well-documented system implementations.

1.9 THE STRUCTURED SYSTEM ANALYSIS AND DESIGN METHODOLOGY

Structured system analysis and design is a structured methodology appropriate for the analysis and design of systems which exist in a well-structured environment. It is less appropriate for ill-structured environments; these must be analysed and transformed into well-structured environments.

The differentiation of well-structured from ill-structured environments can be undertaken by assessing a number of key parameters, some of which are listed in Table 1.1. Well-structured environments are definable and sustainable whereas ill-structured environments are generally unsustainable.

Table 1.1

Parameter	Well-structured	Ill-structured
Objectives	Realistic, clear, consistent	Unrealistic, vague, inconsistent
Problem areas	Known, relevant	Vague, not recognised
Requirements	Consistent, useful	Intuitive
Communication	Effective, reliable	Uncertain, unreliable
Attitudes	Flexible, co-operative	Obstructive

Most environments will exhibit some well-structured and some ill-structured characteristics. It is essential only to consider the use of structured systems analysis and design in an environment which is predominantly well-structured. Other methodologies, such as the soft systems methodology, enable the better structuring and understanding of ill-structured situations.

Structured systems analysis and design is divided into three stages, shown in Figure 1.6. In this way a stage represents the activities of a

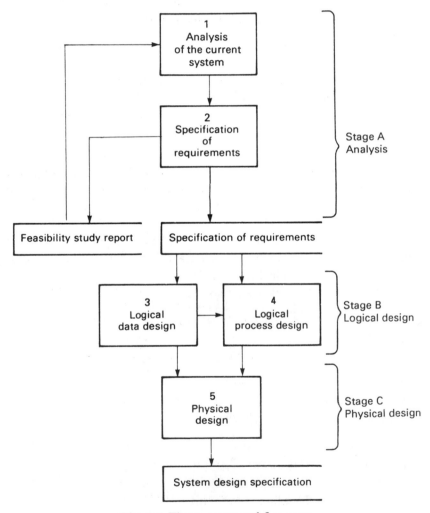

Fig. 1.6 Three stages and five parts

well-defined section of the development cycle. Each stage has its own carefully designed set of objectives. These are:

Stage A   Analysis
Part 1   Analysis of the current system
- to construct a logical model of the current system
- to document the problems with the current system and the requirements for the new system.

Part 2   Specification of requirements
- to construct a business model of the required system, together with detailed documentation.

Stage B   Logical design
Part 3   Logical data design
- to complete a detailed logical data design.

Part 4   Logical process design
- to complete a set of detailed logical process designs.

Stage C   Physical design
Part 5   Physical design
- to translate the logical data design into the file or database specification and the logical process designs into program specifications.

Each stage comprises several parts and tasks. The tasks provide a structured approach leading to the achievement of the stage and part objectives.

Techniques such as data flow diagrams are used to provide a mechanism for the completion of tasks. Data flow diagrams are used for the completion of specific tasks within parts 1, 2 and 5. A variety of techniques are chosen according to appropriateness to the task; structured systems analysis and design is not technique dependent.

1.9.1 Structured systems analysis and design in the systems development cycle

Structured systems analysis and design is a methodology for systems analysis and design. Figure 1.7 shows its coverage of the systems development cycle. It commences by organising the investigation notes and concludes by producing a detailed design specification for the implementors.

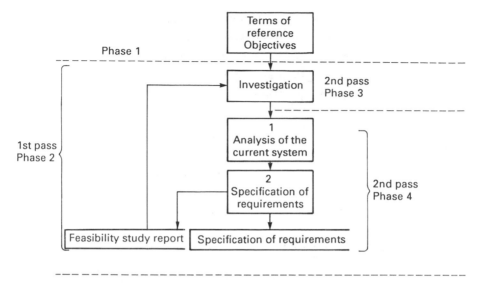

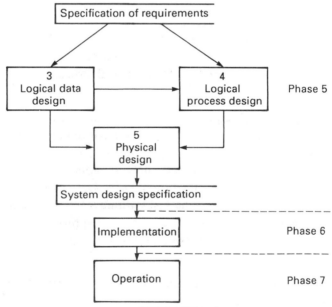

Fig. 1.7 Structured systems analysis and design in the systems development cycle.

Structured systems analysis and design does not include stages for investigation or implementation. The investigation phase of a project relies on techniques common to many business situations, for example, interviewing. Skills in the application of investigation techniques should be possessed by all systems analysts; structured systems analysis and design may commence following any detailed investigation.

Implementation commences with programming and program testing. Structured programming existed well before structured analysis and design, and many organisations have adopted it as an installation standard. Structured systems analysis and design, therefore, completes a total structured approach. Stage C is tailored to a user's environment to ensure that the structured design interfaces with existing programming methods

1.9.2 The benefits of structured systems analysis and design

Many users of structured systems analysis and design have demonstrated that it does not increase the cost or timescale of development projects. Indeed, the experience has shown that there exists great potential to shorten the timescale and decrease the cost of development projects by increasing the chances of meeting the user's exact requirements first time.

The meeting of the user's requirements, first time, is perhaps the greatest benefit to be accrued from a structured approach. This can be attributed very largely to the methodology's involvement of the user throughout the development process.

The format of the documentation makes the system specification understandable by the user. Techniques, such as data flow diagrams, allow users to make their own decisions regarding information requirements and business options without becoming involved in the design technical detail.

A structured approach forces the development team to consider the detail early in the project to ensure that the specification of requirements and the design specification are correct. Many more analysis and design decisions are forced into the earlier stages. This contrasts with many methods of working whereby the true costs of analysis and design are hidden because many decisions taken during the analysis and design stages are based on insufficient knowledge and are corrected during programming and implementation. The costs and timescales associated with the programming and implementation are inflated by this correction factor which has been shown to be up to 100 per cent of the implementation costs and timescales.

The final benefit is realised after implementation: structured systems analysis and design includes detailed documentation as part of the methodology. This ensures that all documentation is completed as part of the project, whereas documentation following analysis and design is rarely completed fully. The documentation provides a means of reducing the costs and timescales of system maintenance and modification.

A structured approach, therefore, brings benefit to many phases of the system development process, from initial analysis and design through to implementation, maintenance and modification during live running.

The major benefit must be restated: it enables users to obtain the systems they require.

Chapter 2
The Techniques Used in Structured
Systems Analysis and Design

2.1 INTRODUCTION

Structured systems analysis and design is not based on any one
technique but is a framework comprising three stages and five parts.

Each part is designed to achieve a specific objective or set of
objectives. The achievement of these objectives is engineered by using
a set of tasks within each part. Each task is based upon the most
appropriate technique for the task. A part may therefore use several
techniques to achieve its objective and, conversely, a technique may be
used during more than one part.

Chapter 3 describes in detail the stages, parts and tasks within
structured systems analysis and design, while this chapter introduces the
techniques. A set of techniques and related documentation that users
can easily understand form the basis of the methodology.

The prime techniques are:

(1) Data flow diagrams which show the boundary of the system and its
    relationship to the external world. They also show the functions,
    data stores, input and output for the system.
(2) Entity models, which show the data structures and data rela-
    tionships for the system. Groups of data structures or entities will
    eventually form the data stores on the final data flow diagrams.
(3) Entity life histories, which show how each entity is affected by
    system functions and provide a dynamic view of a system.
(4) Normalisation, which transforms complex data structures into
    simple lists, is used to build entity models bottom up from the input
    and output data structures.
(5) Process outlines, which specify the operations necessary to process a
    transaction in the system. These lead towards the production of
    program specifications.
(6) Physical design control, which provides sets of rules for transform-
    ing logical specifications into physical specifications. A different set
    of rules is required for each hardware and software combination.

An improvement in the quality of developed systems is one of the claimed benefits of structured systems analysis and design; it uses a number of techniques to assist in this. Cross referencing is a technique used in several parts, to ensure the completeness and accuracy of the models at that stage. Data store entity cross references are built during part 1 and entity function cross references are built during part 2.

User reviews, in the form of walkthroughs for quality control purposes as well as for monitoring project progress, are a formal part of the methodology.

The final technique is related to documentation. The methodology includes an integrated set of documents which support the methodology. Documentation is very much a part of the methodology, not a task which is undertaken when the project is completed.

2.2 DATA FLOW DIAGRAMS

Data flow diagrams provide a view of the system that is understandable to the user: a user's view. The user in any system produces output, very often in the form of new or amended documents, based upon the flow of data into the function the user performs. The user, therefore, observes data flow and is able to describe in detail this view of the system. Data flow diagrams are used in four forms – current physical, current logical, required logical and required physical – during stage A.

The user's view of the information system is modelled by data flow diagrams. They are system models against which analysts can test their understanding by first constructing the model and secondly by performing a structured walkthrough of the model with the user. Data flow diagrams are a simple graphical representation of data flow, data storage and functions which users readily accept. Many users quickly learn sufficient about them so that they can contribute directly to their construction.

The various forms of data flow diagrams represent models of the system during its various stages of refinement:

Current physical   How the existing system operates
Current logical   What the current system accomplishes
Required logical   What the new system is required to accomplish
Required physical   How the required system will be implemented.

Figure 2.1 shows a current physical data flow diagram, and Figure 2.2

DATA FLOW DIAGRAM

SYSTEM: CHAPTER 2	DATE:
AUTHOR: G. CUTTS	PAGE: 1 of 2

LEVEL: 1	CURRENT/~~REQ.~~	PHYS./~~LOGICAL~~

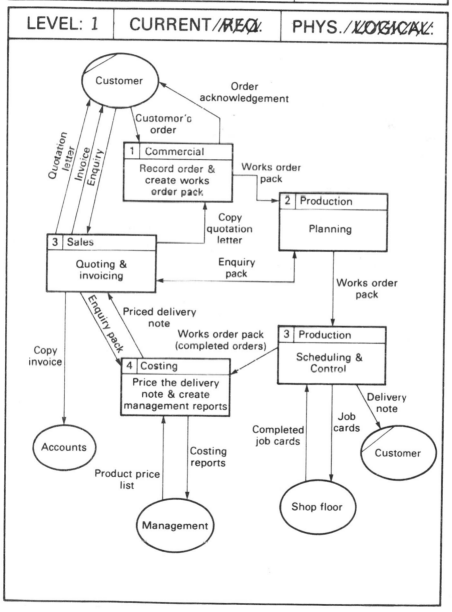

Fig. 2.1

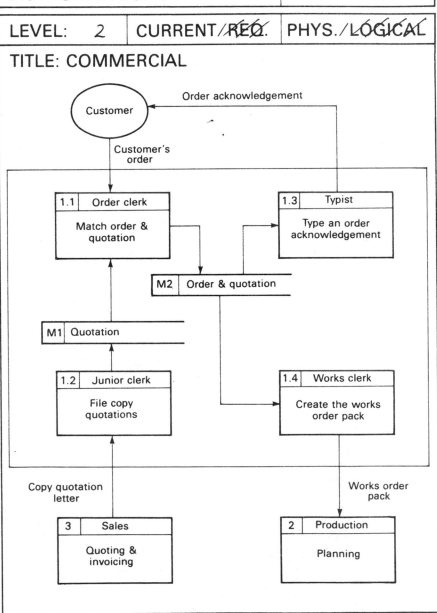

DATA FLOW DIAGRAM

SYSTEM: CHAPTER 2

DATE:

AUTHOR: G. CUTTS

PAGE: 2 of 2

LEVEL: 2   CURRENT/~~REQ.~~ PHYS./~~LOGICAL~~

TITLE: COMMERCIAL

Fig. 2.2

shows an explosion of function 1 from Figure 2.1. Data flow diagrams provide an excellent graphical descriptive tool that is easy for users to understand. They provide a clear view of the system functions and the system boundary and are therefore one of the techniques used to assist analyst–user communication.

The representation of exceptions and errors on data flow diagrams makes them over-complicated, losing the advantage of clarity. In addition, data flow diagrams can become very complex with many levels being required to express very detailed functions. This complexity can lead to omissions and errors on the diagram. Additionally, a methodology based solely on data flow diagrams would not include the structure of data. Data flow diagrams should, therefore, be just one of a set of complementary techniques.

2.3 ENTITY MODELS

Entity models provide a system view of the data structures and data relationships within the system. All systems possess an underlying generic entity model which remains fairly static in time. The entity model reflects the logic of the system data, not the physical implementation.

There are many examples where different departments of the same organisation perform identical functions in very different ways. Their respective current physical data flow diagrams would reflect the differences. Their entity models would be very similar, since the data and data relationships required to perform the function are the same.

Entity models are logical not physical; they represent logical groups of data, called entities, and the relationship between the entities. Customer and order are examples of entities; places (i.e. a customer places an order) is an example of a relationship between the entity customer and the entity order. Figure 2.3 shows an entity model.

Entity models are used in parts 1, 2 and 3. They progress from early attempts to understand the underlying data structures and relationships through to the construction of the final entity model.

Entity models provide an excellent graphical representation of the generic data structures and relationships. They provide a clear view of the logical structure of data within the boundary of interest and allow the analyst to model the data without considering its physical form. Entity modelling provides a system view independent of current

ENTITY MODEL

SYSTEM: CHAPTER 2 · EXAMPLE	DATE:
AUTHOR: G. CUTTS	PAGE: 1 of 1

VERSION: 1

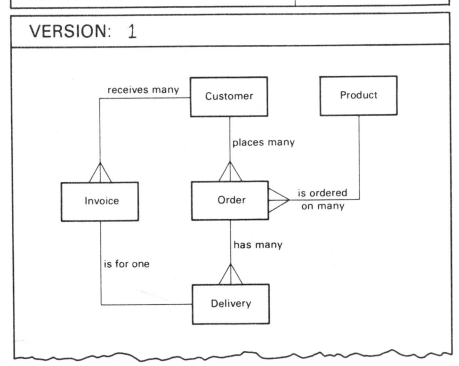

Entity	*Relationship*	*Entity*
Customer	receives many	Invoice
Customer	places many	Order
Product	is ordered on many	Order
Invoice	is for one	Delivery
Order	has many	Delivery

Fig. 2.3

processing; it is a system-wide view not a functionally decomposed view.

The technique of entity modelling complements data flow diagrams, providing the missing system view of the data structures and relationships.

2.4 CROSS REFERENCE

The technique of cross referencing one set of objects to another is used during stage A to provide tests of completeness and accuracy.

The open ended rectangles on data flow diagrams represent data stores. Each data store comprises part of an entity, a complete entity, a number of entities or several parts and complete entities. It is possible to cross reference the data stores from a set of data flow diagrams with the entity model. Figure 2.4 shows a cross reference between a set of current physical data stores and the entity model for the same system.

If a current physical data store cannot be referenced to one or more entities, then either the entity model is incomplete or the physical data is not required within the system. If an entity from the entity model is not cross referenced, then the entity may not be stored in the current system.

Errors may have occurred during the construction of either model. If the construction of a cross reference highlights problem areas, it serves as an early proof mechanism on the completeness and accuracy of the data flow diagrams and the entity model.

The cross reference also provides an early insight into the areas of duplicated information within the current physical system. Duplication of data is often necessary to avoid lengthy access times but nearly always leads to inconsistencies. Figure 2.4 shows that data on orders is held on three physical files: the order file, the stock file and the accounts file.

2.5 ENTITY FUNCTION MATRIX

The construction of an entity function matrix is a technique which leads to the construction of entity life histories and logical process outlines. The technique is used to construct a matrix in which the entities form the rows and the functions form the columns. The rows are formed by listing all the entities from the entity model and the columns are formed

DATA STORE/ENTITY X REF.

SYSTEM: *CHAPTER 2 · EXAMPLE*	DATE:
AUTHOR: *G. CUTTS*	PAGE: *1* of *1*

PHYSICAL/~~LOGICAL~~

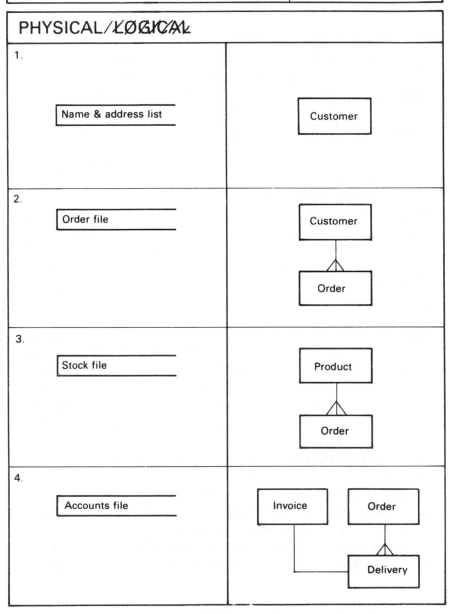

Fig. 2.4

ENTITY/FUNCTION MATRIX

SYSTEM: CHAPTER 2 EXAMPLE	DATE:
AUTHOR: G. CUTTS	PAGE: I of I

Function name / Entity name	1.1 Order entry	1.2 Produce order acknowledgement	1.3 Record delivery	1.4 Produce invoice	2.1 Create new customer account							
Customer												
Order												
Product												
Delivery												
Invoice												

Fig. 2.5

by listing all the functions on the data flow diagrams. Figure 2.5 shows an entity function matrix without any entries.

Matrix entries comprise the single, or combinations of, letters I,R,M,D,A. I indicates that an entity occurrence is inserted by the function; R indicates that an entity occurrence is read by the function; M indicates that an entity occurrence is modified by the function; D indicates that an entity occurrence is deleted by the function; and A indicates that an entity occurrence is archived by the function. The entries into the matrix can be made by reference to the data flow

ENTITY/FUNCTION MATRIX

SYSTEM: CHAPTER 2 EXAMPLE	DATE:
AUTHOR: G. CUTTS	PAGE: 1 of 1

Entity name \ Function name	1.1 Order entry	1.2 Produce order acknowledgement	1.3 Record delivery	1.4 Produce invoice	2.1 Create new customer accounts						
Customer	R	R			I						
Order	I	M	M	A							
Product			M								
Delivery			I	D							
Invoice				I							

Fig. 2.6

diagrams and the data store entity cross reference. Figure 2.6 shows an example of a completed entity function matrix.

The matrix acts as a further proving technique. It is easy to identify that the customer entity needs functions to modify and delete the entity; the product entity needs a maintenance function; the invoice entity is inserted but never accessed; and the delivery entity is simply inserted, then deleted. These potential anomalies require further investigation with satisfactory reason being found for them, or with action being taken to eliminate them before continuation of the project.

Reading across a row of the matrix gives the entity life history. The entity order is inserted by function 1.1, order entry; modified by functions 1.2, produce order acknowledgement, and 1.3, record delivery; and eventually archived by function 1.4, produce invoice. Reading down a column gives the processing that a function has to carry out. Function 1.3, record delivery, modifies the order and product entities, and inserts the delivery entity.

The entity function matrix, therefore, provides a technique for plotting the effect of functions on entities. It also provides a way forward into entity life histories and process outlines.

Cross references such as the data store entity and the entity function provide a technique for proving the models, therefore improving the quality of the final system.

2.6 ENTITY LIFE HISTORY

Each row of the entity function matrix shows how an entity is affected by functions. It does not show the sequence of functions nor does it show when it is valid to carry out a function. Further the matrix shows the normal functions which affect the entity. The entity life history technique provides a graphical representation which shows the sequence of functions and when it is acceptable to carry out a function. Entity life histories also show abnormal functions. Figure 2.7 shows a simple entity life history with no abnormal functions.

The squares represent functions, and the circles show the current status of the entity occurrence. The function, make payment, 4.1, modifies the entity customer account. It is only valid to make payments when the entity occurrence has a status of two. The status of the entity occurrence should be re-set to two, in this instance, on completion of successful processing. The function, make final payment, 4.2, is only valid if the status is two; the 'set to' status, in this case, is three. The only valid function, when the status is three, is close account. Each function possesses, for each entity it affects, a 'valid previous' status and a 'set to' status.

Entity life histories allow the analyst to concentrate on a particular entity. In this case, perhaps the entity life history should be amended to allow subsequent mortgage agreements. Figure 2.8 shows an amended entity life history for the entity, customer account.

ENTITY LIFE HISTORY

SYSTEM: CHAPTER 2 EXAMPLE	DATE:
AUTHOR: G. CUTTS	PAGE: 1 of 1

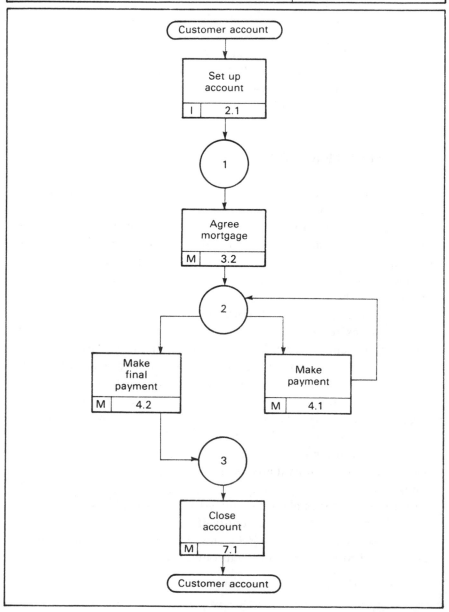

Fig. 2.7

ENTITY LIFE HISTORY

SYSTEM: CHAPTER 2 EXAMPLE | DATE:

AUTHOR: G. CUTTS | PAGE: 1 of 1

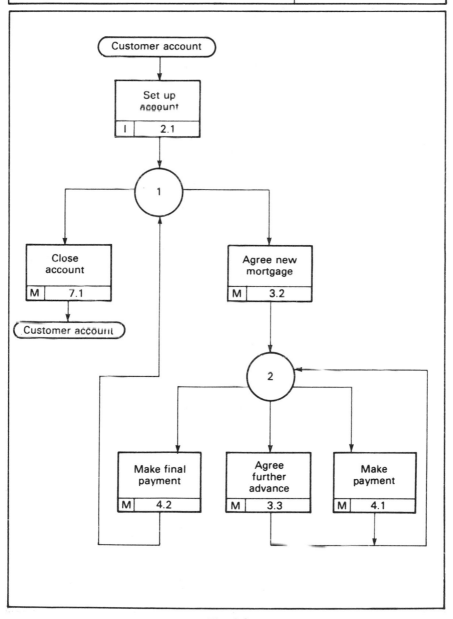

Fig. 2.8

Customer Number	Name	ISBN	Title	Author	Price	Qty. Ordered
A471	J. Allen	0-13-47917-3	Data Processing	A. Downs	15.00	10
		0-12-27947-1	Program Design	J. Smith	12.00	5
		0-17-11001-2	IKBS	F. Williams	22.00	1
J419	J. Jones	0-12-27947-1	Program Design	J. Smith	12.00	7
K579	A. Keith	0-17-11001-2	IKBS	F. Williams	22.00	2
		0-21-100001-3	Novel Architecture	N. Wills	15.00	3

Fig. 2.9

2.7 NORMALISATION

Normalisation is the technique used to produce data structures in third normal form, from input and output specifications. Input and output data structures are progressively transformed from UNF (unnormalised form) through 1NF and 2NF (first and second normal forms) to 3NF (third normal form).

Normalisation is a technique for transforming complex data structures into simple tables. The tables form the building blocks for data design; the methodology uses the tables to create entities and an entity model.

Figure 2.9 shows a complex data structure representing orders for books. Much of the data relating to books is duplicated in the data structure, which is unnormalised. Note also that with this data structure, customer details and book details cannot be stored unless an order exists. Normalisation would result in three simple tables, shown in Figure 2.10; these tables are in third normal form. First and second normal form represent steps towards the target, third normal form.

Customer Table

No.	Name
A471	J. Allen
J419	J. Jones
K579	A. Keith

Book Table

ISBN	Title	Author	Price
0-13-47917-3	Data Processing	A. Downs	15.00
0-12-27947-1	Program Design	J. Smith	12.00
0-17-11001-2	IKBS	F. Williams	22.00
0-21-100001-3	Novel Architecture	N. Wills	15.00

Order Table

Customer No.	ISBN	Qty. Ordered
A471	0-13-47917-3	10
A471	0-12-27947-1	5
A471	0-17-11001-2	1
J419	0-12-27947-1	7
K579	0-17-11001-2	2
K579	0-21-100001-3	3

Fig. 2.10

Normalisation results in the discovery of entities and entity descriptions, bottom up, from detailed input and output descriptions.

2.8 PROCESS OUTLINES

A process outline collects together all the operations that are necessary for a process to execute. The entities affected by the processing can be established from the entity function matrix. The previous status of the entity occurrence, for which the function is valid, can be established from the entity life histories. The entity life history will also provide the 'set to' status on the successful completion of the function.

A skeletal process outline comprising the entity name, the effect on the entity occurrence, the 'valid previous' status and the 'set to' status can easily be constructed. An example is shown in Figure 2.11.

Create invoice		Status	
Entity	Effect	Valid previous	Set to
Order line	M	3	3
Delivery	M	2	3
Invoice	I	—	1

Fig. 2.11 Skeletal process outline for create invoice

The skeletal process outline can now be enhanced to include a description of the processing for each operation required to complete the process. Each process outline will include one or more function boxes from the data flow diagrams. A completed process outline is shown in Figure 2.12.

2.9 PHYSICAL DESIGN CONTROL

Physical design control is not a single technique, but a number of rules for converting the logical data design and logical process outlines into physical specifications. The result of physical design control is a database or file specification and a series of program specifications.

The conversion rules will vary according to the target hardware and software.

Database or file specifications are enhanced or modified entity

LOGICAL PROCESS OUTLINE

SYSTEM: CHAPTER 2 EXAMPLE	DATE:
AUTHOR: G. CUTTS	PAGE: 1 of 1

EVENT NAME: Create Invoice	VOLUME:

BRIEF DESCRIPTION: Invoices are created from the input of actual quantity despatched taken from a copy of the delivery note.

Op. No.	Entity Name	Eff.	Status Ind. prev	set	Description Narrative	Ref.	I/O Ref.	Err. Ref.
1					Read input delivery details. Validate the input fields as described in VIO. If errors are detected follow the procedure described in E2 and the details in format O2.	VIO	O2	E2
2	Order line	M	3	3	Find the relevant order line If not found reject the transaction, see O3, E3. If found but status error, see EI Add 1 to the number of invoices field.		O3	E3 EI
3	Del-ivery	M	2	3	Find the relevant delivery If not found reject the transaction, see O3, E3 If found but status error, see EI Modify the quantity to the quantity input.		O3	E3 EI
4	Invoice	I	—	1	Insert invoice			

Fig. 2.12

models. The entity model is enhanced to show access points and access methods, then modified to satisfy the constraints of the database or file management software available.

Physical program specifications are enhanced process outlines. Several process outlines may be merged to create one program specification. Enhancements to the process outline include the details of how each entity is accessed via the database implementation.

2.10 WALKTHROUGH

A walkthrough is a meeting of all interested people to discuss the content of the project. It is not a progress review.

2.11 STAGES, PARTS AND TECHNIQUES CROSS REFERENCE

Structured systems analysis and design comprises three stages and five parts; the techniques used within each stage and part are:

Stage A   Analysis
Part 1   Analysis of the current system
- data flow diagrams
- entity models
- cross reference
- walkthroughs and documentation
Part 2   Specification of requirements
- data flow diagrams
- entity models
- cross reference
- entity life histories
- walkthroughs and documentation

Stage B   Logical design
Part 3   Logical data design
- normalisation
- entity models
- walkthroughs and documentation
Part 4   Logical process design
- process outlines
- walkthroughs and documentation

Stage C Physical design
Part 5 Physical design
- physical design control
- walkthroughs and documentation

2.12 SUMMARY

The objective of the structured systems analysis and design methodology is to provide a framework for systems analysis and design. The framework comprises several techniques, and a system of documentation.

The techniques have been chosen for their applicability to the stage, part and task as well as for being usable and teachable. The major techniques are data flow diagrams, for representing information flows; entity models, for representing data structures and data relationships; entity life histories, for representing the effect of time; normalisation, for building third normal form data structures; process outlines, for specifying detailed process logic; and physical design control, for establishing optimum physical database, file and program specifications.

The techniques are now fitted into the detailed framework of stages, parts and tasks in Chapter 3.

Chapter 3
Structured Systems Analysis and Design: Stages, Parts and Tasks

3.1 INTRODUCTION

The terms of reference are the starting point for structured systems analysis and design projects. They state the business objectives for the proposed development and the resources available for it. In many cases, a feasibility study report will have been completed which will form a second input.

The outputs are a set of program specifications, a database specification, an implementation plan, a user manual and an operations manual. Structured systems analysis can also be used to produce feasibility study reports.

Figure 3.1 shows the methodology's inputs and outputs. The stages and parts have been constructed so that each has a clear objective with limited scope. Each part has a defined set of inputs and outputs which provide a clear, clean interface between the parts.

Figure 3.2 shows the five parts of the methodology, its inputs and outputs, and the input and output of each part. The division into five parts provides a high level project management tool. Since each part is self-contained and has a precise objective and a set of outputs, management monitoring of completion is relatively easy. A walk-through should be undertaken at the end of each part to monitor both progress and quality. The parts could be used to create project milestones.

The inputs and outputs from each part are:

Stage A   Analysis
Part 1   Analysis of the current system
  - Input   Terms of reference
               Feasibility study report (optional)
  - Output   Current logical data flow diagrams
               Current entity model
               Problems and requirements list

42

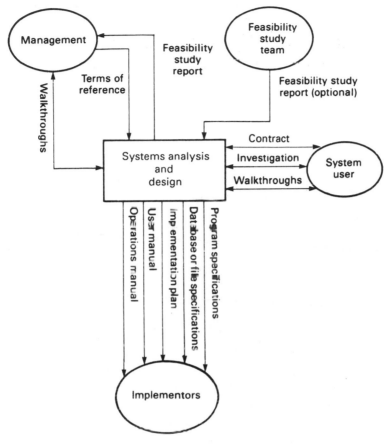

Fig. 3.1

Part 2   Specification of requirements
   • Input   Current logical data flow diagrams
                Current entity model
                Problems and requirements list
   • Output   Required entity model
                Required physical data flow diagram
                Design constraints
                Entity descriptions
                Input and output descriptions
                Data dictionary
                On-line dialogue specification

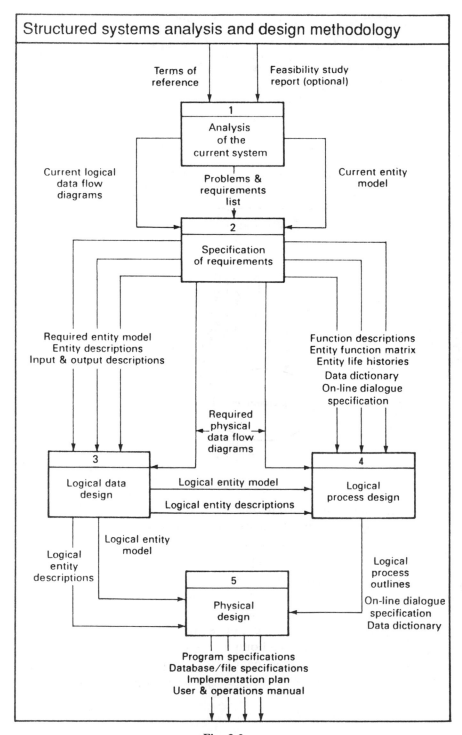

Fig. 3.2

                Function descriptions
                Entity function matrix
                Entity life histories

Stage B   Logical design

Part 3   Logical design data
- Input   Required physical data flow diagrams
                Required entity model
                Entity descriptions
                Input and output descriptions
- Output   Logical entity model
                Logical entity descriptions

Part 4   Logical process design
- Input   Required physical data flow diagrams
                Logical entity model
                Logical entity descriptions
                Input and output descriptions
                On-line dialogue specification
                Function descriptions
                Entity function matrix
                Entity life histories
- Output   Logical process outlines

Stage C   Physical design

Part 5   Physical design
- Input   Logical process outlines
                Logical entity model
                Logical entity descriptions
                Data dictionary
- Output   Program specifications
                Database or file specifications
                An implementation plan
                A user manual
                An operations manual

3.2 SYSTEM MANUAL

The inputs shown for each part form the major source of information. One of the major benefits of a structured approach is the gradual completion of the system documentation, part by part. The system manual or system documentation is not a specific deliverable from any of the parts, but from all of the parts. The system manual is available throughout the life of the system and therefore forms a secondary input to each part.

3.3 STRUCTURED SYSTEMS ANALYSIS AND DESIGN: PARTS

3.3.1 Part 1: Analysis of the current system

Part 1 uses the terms of reference and, optionally, the feasibility study report to produce a logical model of the current system. The logical model comprises a set of current logical data flow diagrams and a current entity model. Part 1 commences with a planned investigation of the current system, the results of which are used to build the data flow diagrams and entity model. During the building of the models and during the investigation, problems with the current system and requirements for the new system will be identified. A list of problems and requirements forms the third deliverable item from part 1, along with the current logical data flow diagrams and the current entity model.

The objectives are to obtain a thorough understanding of the functions and data within the current system; to build logical models of what data flows, data stores, functions, data structures and data relationships exist; to identify problems with the current system; to record requirements for the new system; and to agree the logical models, problems and requirements with the user.

3.3.2 Part 2: Specification of requirements

Part 2 can be divided into three sections. Section one uses the current models as input and optimises them before creating the required models. The required models reflect solutions to the problems documented, as well as building into the models the requirements for the new system. The required logical data flow diagrams may be created from the current logical data flow diagrams, and the required entity model may be created from the current entity model.

Logical diagrams model what is required of the system; physical diagrams model how the requirements are to be provided.

The objective of section one is to select a particular business specification for further development. The selection, from a short list of business options developed from the required models, should be made by the user, assisted by the project team. Each option will have a different impact on the problem and the requirements, and on the user; it will also have different costs and implementation times. It is these considerations, along with technical guidance from the project team, which will lead the user to a final selection.

Section two provides the detailed documentation for the business specification. Entity descriptions and input and output descriptions are documented as attribute lists. The input and output descriptions form part of the detailed documentation of the data flow diagrams along with the entity descriptions, which effectively document the data stores. To complete the documentation of the data flow diagrams, function descriptions are required. Each function which is not further decomposed should be documented. Finally, the on-line dialogue is specified and the data dictionary is produced.

Section three provides a detailed third view of the system – a view, in addition to the static views provided by data flow diagrams and entity models, which models the effects of functions on entities in time. This view is dynamic and provides early proving of the model building tasks.

The objective of stage A is to provide a clear, concise, unambiguous, complete set of documents which specifies the required system in terms that the user can understand. A walkthrough at the end of stage A must reach agreement on the new system.

3.3.3 Systems analysis

The specification of the required system marks the conclusion of systems analysis, a major milestone in any project. Stage A is the analysis stage, the objective being to produce a detailed specification of requirements. For this reason, a major milestone walkthrough should be planned for the end of part 2.

Figure 3.3 shows the specification of requirements as a major interface in the methodology.

Structured systems analysis and design may be used in two modes up to part 2, single pass and two pass analysis. With two pass analysis, stage A is used as a methodology to provide an overview of the system and to provide the information necessary for the production of the feasibility study report.

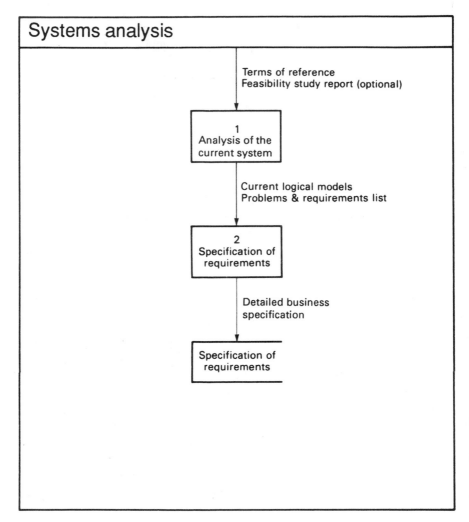

Fig. 3.3

Stage A appears in the methodology twice. The full two pass methodology for systems analysis is shown in Figure 3.4.

3.3.4 Part 3: Logical data design

Part 3 uses the input and output descriptions to produce data structures in third normal form. An entity model and entity descriptions are

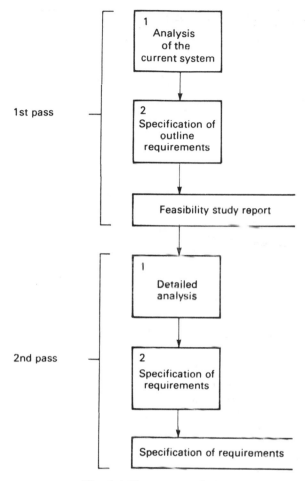

Fig. 3.4 Two-pass analysis

created from the data structures. The entity models and entity descriptions from stage A are compared with those just produced, differences are resolved by reference to the requirements and the user, and a set of logical entity descriptions and a logical entity model are delivered to parts 4 and 5.

The objective of part 3 is to ensure that the data structures and data relationships are fully described and understood. This part produces entity descriptions and the entity model bottom up, whereas during part 2 they were produced top down. The two approaches ensure that a high-quality logical entity model and logical entity descriptions are delivered to parts 4 and 5.

3.3.5 Part 4: Logical process design

The first task in part 4 is to catalogue the functions from the required physical data flow diagrams. The functions are catalogued according to their type of processing (on-line, batch), the timescale of processing (daily, weekly) and the access requirements (sales ledger, order file). Each process will comprise one or more data flow diagram functions.

For each logical process, a process outline is created which describes the operations necessary to execute the process. This logical process outline represents a logical program or module specification, the output from part 4 passed to part 5.

The objective of part 4 is to group the functions into logical processes according to processing requirements, and to provide detailed descriptions of the logical processes.

3.3.6 Part 5: Physical design

In physical design, the logical entity model and logical entity descriptions are transformed into a database or file specification by the application of rules specifically designed to reflect the target hardware and software. The database or file specifications are then used, along with a second set of rules, to transform the logical process outlines into program specifications. Optimisation and tuning of the specifications is considered before production of the final database or file specifications and the final program specifications.

The system specification is now complete and part 5 can conclude by producing an implementation plan covering programming, conversion etc., an operations manual and a user manual.

3.3.7 Systems design

Systems design includes stages B and C. Systems design is the transformation of the specification of requirements, first into a detailed logical specification and secondly into a detailed physical specification.

There are two major milestones during systems design. These are at the end of stage B, when a complete logical design is available for review, and at the end of stage C, when the complete physical design is available for review. Note that the target hardware and software are not considered until stage C; detailed knowledge of the hardware and software is not required to be able to complete the logical design. System designers, therefore, need not be technical experts to be able to undertake logical design. Indeed, systems design can proceed up to detailed logical design followed by several stage Cs, each targeted on a different hardware and software system.

Systems design is clearly separable into logical design and physical design.

3.4 STRUCTURED SYSTEMS ANALYSIS AND DESIGN: TASKS

Each part possesses clearly-defined objectives, a set of inputs and a set of outputs. The inputs are transformed into the outputs to meet the objectives of the part by a set of tasks. Each task within each part possesses a clearly defined objective: a sub-objective of the part's objectives. Similarly, each task possesses a set of inputs and a set of outputs.

The stages are labelled A, B, C; the parts are numbered 1 to 5; the tasks are numbered within each part. Task 5 in part 1 is therefore numbered 1.5. The user of the methodology can clearly understand the stage, part and task being undertaken, the inputs, the outputs, the objective and the techniques to be used for the task. Structured systems analysis and design provides a structured approach to systems development.

This chapter will now document each task, its objective, its inputs and its outputs, a list of techniques to be used and a brief description of the task. Chapters 4 to 9 will describe each part in detail.

3.4.1 Part 1: Analysis of the current system: tasks

There are six tasks, shown in Figure 3.5.

Part 1, Task 1.1: Investigate the current system
Objective To document the functions and data within the current system.
Input Terms of reference; feasibility study report; interview notes;

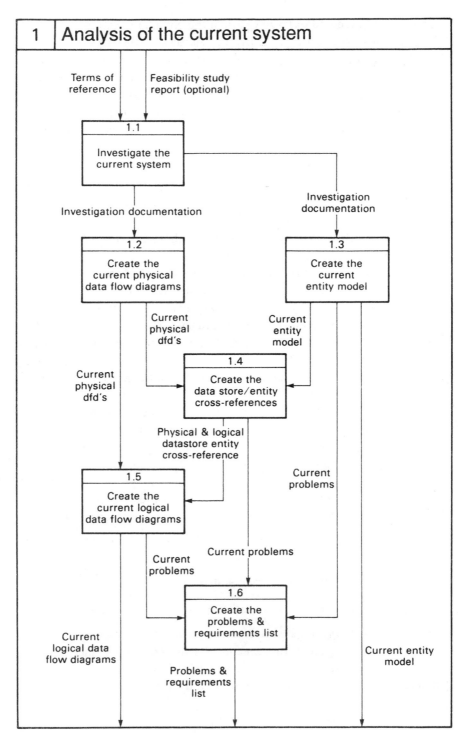

Fig. 3.5

results of investigations of current files; procedures etc.; results of questionnaires etc.

Output A detailed set of investigation notes/documentation.

Technique Any investigation technique or techniques available, such as interviews, questionnaires, current documentation etc.

Description This task is common to all methodologies. It uses the standard skills all systems analysts possess and seeks to gain a thorough understanding of the current system.

Part 1, Task 1.2: Create the current physical data flow diagrams

Objective To model the physical functions, data flows, data stores and external entities associated with the current system. The model represents how the current system operates.

Input Investigation documentation; terms of reference

Output A set of physical data flow diagrams.

Technique Data flow diagrams; walkthroughs.

Description This task provides a graphical representation of the current system which models the user's view of the system. The user's view comprises the document flow, the files and stores of data, the providers and recipients of documents and the functions carried out.

The current physical model will model all of the problems and anomalies of the current system since it should be a faithful representation of current practice.

Part 1, Task 1.3: Create the current entity model

Objective To model the logical data structures (entities) and entity relationships required to support the current system.

Input Investigation documentation.

Output Current entity model.

Technique Entity modelling.

Description This task provides a graphical representation of the current system which models the system view. The system view comprises the entities and relationships necessary to support the processing requirements of the system. The model represents the underlying generic data structures and is a logical model. It does not, therefore, reflect the current files and data stores directly.

Part 1, Task 1.4: Create the data store entity cross reference

Objective To provide a physical data store and a logical data store cross reference which will enable the logicalisation of the data flow diagrams.

Input   Current physical data flow diagrams; current entity model.
Output   Physical data store entity cross reference; logical data store entity cross reference.
Technique   Cross reference.
Description   This task first provides a cross reference between the current physical data stores and the entities from the entity model. It provides a check on the completeness and accuracy of the data flow diagrams and the entity model. It also allows each data flow, to or from a physical data store on the current physical data flow diagram, to be annotated with one or more entity names.

The task also provides a cross reference between current logical data stores and the entities. This is obtained by partitioning the entity model into logical groups.

Part 1, Task 1.5: Create the current logical data flow diagrams
Objective   To model the logical functions, data flows, data stores and external entities associated with the current system. This model represents what the current system accomplishes.
Input   Current physical data flow diagrams; physical data store entity cross reference; logical data store entity cross reference.
Output   Current logical data flow diagrams.
Technique   Data flow diagrams.
Description   This task uses a four-step procedure to transform the physical model into a logical model. The steps include logicalising the data flows, data stores and functions, as well as removing physical time dependencies and physical only functions.

The logical model represents what is accomplished by the current system, not how the current system operates.

Part 1, Task 1.6: Create the problems and requirements list
Objective   To document the problems with the current system and the requirements for the new system.
Input   Terms of reference; investigation documentation; perceived problems and requirements.
Output   The problems and requirements list.
Technique   Itemised formatted list.
Description   The investigation task, particularly during the interview procedure, should reveal the problems with the current system. Requests for new system functions and data are received during the investigation. The problems and requirements are formally documented

by this task. The construction of logical models of the current system also reveals problems and requirements. The problems and requirements list, therefore, documents all the problems with the current system and requirements for the new system discovered during part 1.

Part 1 Review
Objective   To agree with the user the accuracy and completeness of the current system model and the list of problems and requirements.
Input   The current logical data flow diagrams; the current entity model; the problems and requirements list.
Output   An agreement.
Technique   Walkthrough.
Description   This task ensures that the final models of the current system, together with the problems and requirements, are 'signed off' by the user. The walkthrough may result in an iteration of some or all of the tasks within part 1. Part 1 cannot be considered complete until a successful walkthrough with the user has taken place.

3.4.2 Part 2: Specification of requirements: tasks

There are twelve tasks, shown in Figure 3.6.

Part 2, Task 2.1: Create the required logical data flow diagrams
Objective   To model the logical functions, data stores, data flows and external entities associated with the required system.
Input   The current logical data flow diagrams; the problems and requirements lists.
Output   The required logical data flow diagrams.
Technique   Data flow diagrams.
Description   This task generates the required functional model from the current functional model. Solutions are found to the problems generally by providing an enhanced system. The data flow diagrams are, therefore, optimised and enhanced or re-created to solve the problems and to satisfy the new requirements.

Part 2, Task 2.2: Create the required entity model
Objective   To model the logical data structures (entities) and relationships associated with the required system.
Input   The current entity model; the required logical data flow diagrams; the problems and requirements list.

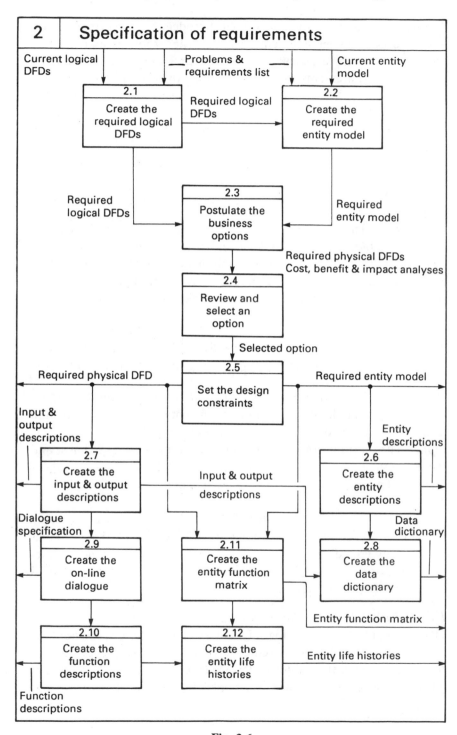

Fig. 3.6

Output   The required entity model.
Technique   Entity modelling.
Description   This task transforms the current entity model into the required entity model. The required logical data flow diagrams may well differ from the current logical data flow diagrams. New data stores and modifications to the existing data stores will be required to solve the problems with the current system and to satisfy the requirements for the new system.

The new data stores and the modifications to existing data stores effectively introduce new entities, new relationships, entity modifications and relationship modifications. These modifications are used to transform the current entity model into the required entity model.

Part 2, Task 2.3: Postulate business options
Objective   To document a number of possible business options for the required system.
Input   The required logical data flow diagrams and the required entity model.
Output   For each option, a set of required physical data flow diagrams, supporting cost benefit and impact analyses, and supporting narrative.
Technique   Data flow diagrams; cost benefit and impact analyses; narrative.
Description   This task identifies a number of implementation or business options represented by physical data flow diagrams. Each option will comprise a set of required physical data flow diagrams supported by a description of the costs, benefits and impact on the user. For example, one option may specify more on-line processing than a second or even third option.

Part 2, Task 2.4: Review and select option
Objective   To select one of the physical business options for continued development.
Input   The required physical data flow diagrams; the cost benefit and impact analyses; supporting narrative.
Output   A set of required physical data flow diagrams.
Technique   Presentation; walkthrough.
Description   Since each of the options presents a different business solution, the user, assisted by the project team, must select one of them for continued development. The project team should prepare a presentation of the business options and be prepared to walkthrough

each one with the user. This task represents a major milestone in stage A.

Part 2, Task 2.5: Set the design constraints
Objective To set further constraints on the design not already documented.
Input The business option specification.
Output The design constraints.
Technique Narrative.
Description This task documents the constraints on the final design that are not apparent from the specification already completed. The constraints include security, privacy, auditing and recovery requirements.

Design objectives may also be set or revised during this task. The design objectives include resource usage, for example direct access space and device utilisation; and performance objectives, for example response times and batch run times.

Part 2, Task 2.6: Document the entity descriptions
Objective To document the attributes for each entity.
Input Required entity model; investigation documentation.
Output A set of entity descriptions.
Technique Standard format document.
Description This task documents, for each entity, the attributes which comprise the entity description. At this stage the entity descriptions are completed top down by allocation of attributes to entities.

Part 2, Task 2.7: Prototype and document the system input and output descriptions
Objective To document the attributes for each input and output.
Input Required physical data flow diagrams.
Output A set of system input descriptions and a set of system output descriptions.
Technique Prototyping; standard format document.
Description This task uses the required physical data flow diagrams to identify those data flows which cross the system boundary. These data flows comprise the system input and output.

This task documents, for each input and each output, the attributes which comprise the data flow. The use of prototyping leads to the specification of screen and report designs as well as confirming the content of the input and output.

Part 2, Task 2.8: Create the data dictionary
Objective   To document all attributes.
Input   The system input and output descriptions; the entity descriptions.
Output   A set of data item definitions.
Technique   Standard format document.
Description   All data items from the input, output and entity descriptions are entered into the data dictionary. This task starts the documentation of the data items for the implementation team.

Part 2, Task 2.9: Prototype and specify the on-line dialogue
Objective   To document the menu and dialogue structure.
Input   The required physical data flow diagrams; the input and output descriptions.
Output   The dialogue specification.
Technique   Prototyping and dialogue design.
Description   Menu, process and other icons are used to specify the dialogue which may be prototyped, free-standing, or linked to the screen designs, to facilitate agreement with the user.

Part 2, Task 2.10: Document the function descriptions
Objective   To describe the processing carried out by each primitive function.
Input   Required physical data flow diagrams.
Output   A set of function descriptions.
Technique   Structured English; pseudo code; decision tables; decision trees; any other descriptive technique available to the analyst and acceptable to the user.
Description   This task uses one or more of the techniques to describe, and therefore document, the detailed processing carried out by each function. Only functions which are not further decomposed, i.e. primitive functions, require to be described.

   Each description should use a technique and style which is understandable by the user.

Part 2, Task 2.11: Create the entity function matrix
Objective   To chart the effect, in time, of the functions on the entities.
Input   Required physical data flow diagrams; required entity model.
Output   The entity function matrix.
Technique   Cross reference.

Description This task creates a matrix or cross reference of the functions against the entities. The functions, which form the columns of the matrix, represent primitive functions from the required physical data flow diagrams. The entities, which form the rows of the matrix, are simply listed from the entity model.

The entities of the matrix may be one or more of the following:

I   the function INSERTS an occurrence of the entity into the database
M   the function MODIFIES an occurrence of the entity in the database
R   the function READS an occurrence of the entity in the database
D   the function DELETES an occurrence of the entity in the database
A   the function ARCHIVES an occurrence of the entity in the database
space   the function has no effect on the entity.

Part 2, Task 2.12: Create the entity life histories
Objective To document, for each entity, the sequence or sequences of functions which affect the entity.
Input The entity function matrix; the required physical data flow diagrams.
Output A set of entity life histories.
Technique Entity life history.
Description The entity life history, drawn for each entity, shows the sequence or sequences of functions which chart the normal processing of an entity from insertion into the database to deletion from the database. In addition, abnormal and error functions are charted. Finally, a status is attributed to each entity. The status defines which function or functions may access and process the entity occurrence.

Part 2 Review
Objective To agree with the user the accuracy and completeness of the specification of requirements.
Input The documents which form the detailed business specification of requirements.
Output An agreement.
Technique Walkthrough.
Description This task ensures that the specification of requirements is fully understood and accepted by the user; this is vital to stage B. A

successful walkthrough of the documents which comprise the specification of requirements must be completed to conclude stage A.

3.4.3 Part 3: Logical data design: tasks

There are six tasks, shown in Figure 3.7.

Part 3, Task 3.1: Select data structures
Objectives To select a set of data structures for task 3.2, normalisation. The set must include sufficient data structures to enable a complete and accurate entity model to be produced while minimising the effort required for the task of normalisation.
Input The required physical data flow diagrams.
Output A selected data structures list.
Technique Walkthrough.
Description All of the data structures, that is, input, output, data flow and data stores, are represented on the required physical data flow diagrams. A set of data structures is selected by walkthrough, according to a number of selection guidelines. The walkthrough should be by members from the analysis team and the design team.

Part 3, Task 3.2: Normalisation
Objective To transform the selected data structures into simpler structures, thus removing various anomalies associated with complex data structures.
Input A set of selected data structures; the data structure descriptions.
Output A set of simple data structures.
Technique Normalisation.
Description Normalisation is a mathematical technique which transforms complex data structures or unnormalised data structures into simple structures in third normal form. The various anomalies associated with unnormalised data structures are gradually removed during the process of normalisation.

Part 3, Task 3.3: Create the entity descriptions
Objective To create a list of entities, and for each a list of attributes.
Input The data structures in third normal form.
Output A set of entity descriptions.
Technique Set of rules.
Descriptions This task simply merges together all of the third normal

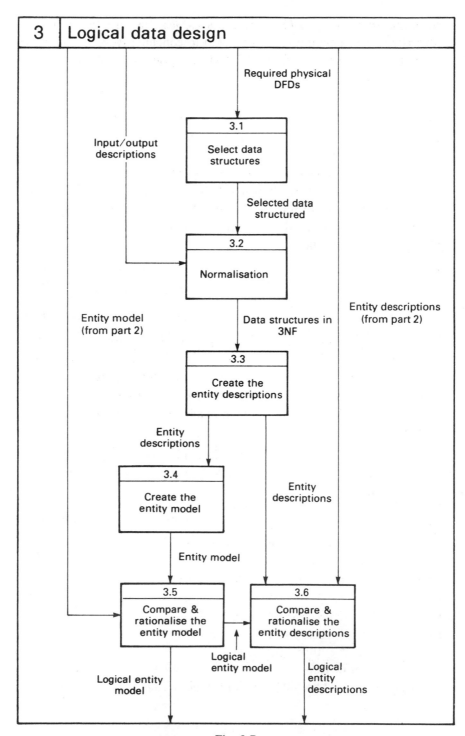

Fig. 3.7

form data structures which have identical keys. The combined lists of attributes form the entity description. The entity itself is uniquely identified by its key, which often provides a pointer to an appropriate entity name.

Part 3, Task 3.4: Create the entity model
Objective   To create an entity model from the entity descriptions.
Input   The entity descriptions obtained during task 3.3.
Output   An entity model.
Technique   The application of a set of prescriptive rules.
Description   Each entity description forms an entity in the entity model. A set of rules is then used to build entity relationships and thus complete the entity model.

Part 3, Task 3.5: Compare and rationalise the entity models
Objective   To produce the final entity model, the logical entity model.
Input   The entity model from part 2; the entity model from part 3.
Output   The logical entity model.
Technique   Inspection.
Description   A series of steps is used in this task to bring together the two entity models. During part 2 the entity model was created by top down analysis and during part 3 it was created bottom up from raw data. This task compares the two, rationalises the difference, often by reference to the user's requirements, to produce the logical entity model.

Part 3, Task 3.6: Compare and rationalise the entity descriptions
Objective   To produce the final set of entity descriptions; the logical entity descriptions.
Input   The entity descriptions from part 2; the entity descriptions from part 3; the logical entity model.
Output   A set of logical entity descriptions.
Technique   Inspection.
Description   The two sets of entity descriptions are compared, any differences are rationalised, and a logical entity description is produced for each entity on the logical entity model.

3.4.4 Part 4: Logical process design: tasks

There are three tasks, shown in Figure 3.8.

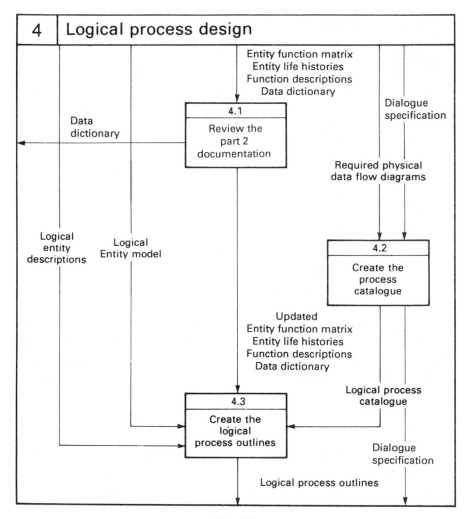

Fig. 3.8

Part 4, Task 4.1: Review the stage A documentation

Objective   To update the entity function matrix, the entity life histories, the function descriptions and the data dictionary as a result of the further development carried out during part 3.

Input   The entity function matrix; the entity life histories; the function descriptions; the data dictionary.

Output   An updated entity function matrix; an updated set of entity life histories; an updated set of function descriptions; an updated data dictionary.

Technique   Inspection or walkthrough.

Description   New entities and relationships and modifications to current ones may well have resulted from the detailed data design tasks during part 3. Other modifications to the stage A specification of requirements, for example modifications to functions, may well have resulted from part 3. The modifications need to be reflected in the entity function matrix, entity life histories, the function descriptions and the data dictionary, before these documents can be used to develop the logical process outlines.

Part 4, Task 4.2: Create the process catalogue

Objective   To group and catalogue the functions by type of processing, by processing frequency and by access requirements.

Input   The required physical data flow diagrams.

Output   The process catalogue.

Technique   Inspection and walkthrough.

Description   This task identifies individual processes in the required system by grouping one or more primitive functions into a single process. Each process will be catalogued under the headings processing mode, on-line/batch; frequency of processing and access requirements. Each process will be for one mode, one frequency and one set of access requirements.

Part 4, Task 4.3: Create the logical process outlines

Objective   To create, for each identified process, a detailed specification of the operations which satisfy the processing requirements.

Input   The updated entity function matrix; the updated entity life histories; the updated function descriptions; the logical entity model; the logical entity descriptions; the logical process catalogue; the data dictionary; the on-line dialogue specification.

Output   A set of logical process outlines.

Technique   Process outlines.

Description   This task expands the set of function descriptions which comprise each process into a detailed, operation-by-operation, description of the process.

The entities affected by the functions within the process are identified from the entity function matrix; the effect on each entity from the entity life history; and the processing required from the function descriptions.

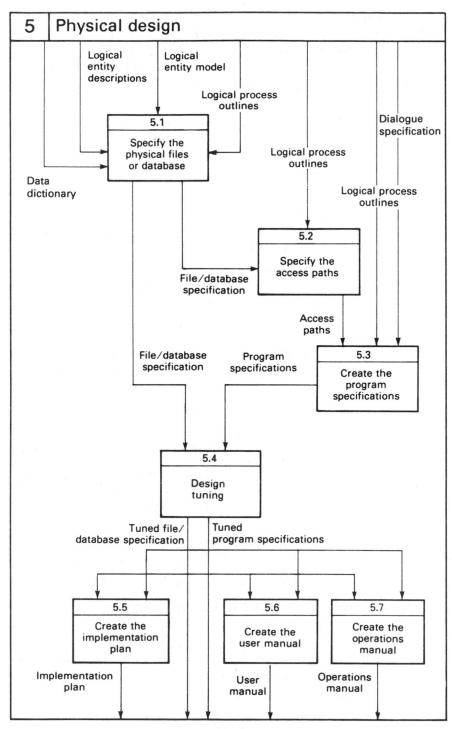

Fig. 3.9

Stage B Review
Objective   To 'sign off' the logical design specification.
Input   The logical entity model; the logical entity descriptions; the logical process outlines.
Output   An agreement.
Technique   Walkthrough.
Description   This task ensures that the logical process outlines can be satisfied by the data design. Each logical process outline should be 'dry run' against the data design to ensure the data design will fully support the processing required.

3.4.5 Part 5: Physical design tasks

There are seven tasks, shown in Figure 3.9.

Part 5, Task 5.1: Create the physical files/database specification
Objective   To transform the logical data design into a physical database specification.
Input   The logical entity model; the logical entity descriptions; the logical process outlines; the data dictionary.
Description   The technique used and the output produced depend on the target hardware and software. For example, if the target software is a network database, then an appropriate technique might be Bachmann diagrams and the expected output a schema definition including detailed sct and record definitions.

Part 5, Task 5.2: Specify the access paths
Objective   To document, for each process outline, the access mechanism for each entity referenced within the process outline.
Input   The logical process outlines; the physical file or database specification.
Output   A set of entity access paths.
Description   The technique is dependent on the hardware and software support available for the physical files or database. The use of the data manipulation language verbs supported by the software is an appropriate method of specifying entity access paths.

Part 5, Task 5.3: Create the program specifications
Objective   To combine the logical process outlines with the access methods to create the program specifications.

Input   The logical process outlines; the access paths.
Output   A set of program specifications.
Description   The technique is dependent on the target hardware, software and installation standards. The output must be acceptable to the installation team. It may be a complete rewrite of the logical process outline and access paths, together with physical screen, report, menu and dialogue designs, or simply the addition of the designs to the process outlines and access path documents.

Part 5, Task 5.4: Design tuning
Objective   To ensure the file or database specification and the program specifications meet the design constraints set during stage A.
Input   The file or database specification; the program specifications.
Output   A revised file or database specification; a revised set of program specifications.
Technique   Walkthrough.
Description   Design tuning involves the estimation of the system's resource usage and performance, the comparison of those estimates against the objectives and the subsequent review of the design and/or the objectives.

Part 5, Task 5.5: Create the implementation plan
Objective   To create a plan for the next phase of the systems development cycle, the implementation.
Input   The system specification.
Output   An implementation plan.
Description   The technique and therefore the specific outputs are installation dependent. For example the development team, by discussion with the implementation team, could hand over a bar chart showing activities and timescales for the completion of the project.

Part 5, Task 5.6: Create the user manual
Objective   To complete the major part of the user manual.
Input   All of the system documentation.
Output   A part-completed user manual.
Technique   Data flow diagrams; narrative.
Description   This task uses the system documentation together with the experience of the development team and the user to complete the major part of the user manual. The manual will be completed and agreed with the user by the implementation team.

Part 5, Task 5.7: Create the operations manual
Objective   To complete the major part of the operations manual.
Input   All the system documentation.
Output   A part-completed operations manual.
Technique   Data flow diagrams; narrative.
Description   This task uses the system documentation, together with the experience of the development team, the user, and the operations department, to partly complete the operations manual. The manual will be completed and agreed with the user and the operations department by the implementation team.

3.5 STRUCTURED SYSTEMS ANALYSIS AND DESIGN, AND PROJECT CONTROL

Structured systems analysis and design is not a project control system; it does, however, provide a structure by which projects may be planned and monitored.

Strategic plans may be created using the system development cycle together with the stages and parts within structured systems analysis and design. Tactical plans may be created using the tasks. Since each stage, part and task has a specific objective or set of objectives with given deliverable items, monitoring completion is made easier. A task is complete when the documented output has been produced and agreed. Similarly, each task may commence only when required inputs are available.

Structured systems analysis and design comprises three stages, five parts, and thirty-four tasks. The techniques used are only necessary to provide an efficient mechanism for completing tasks. Structured systems analysis and design is, therefore, a development methodology, not a collection of techniques.

3.6 STRUCTURED SYSTEMS ANALYSIS AND DESIGN FOR SMALL SYSTEMS

It is feasible to use a cut-down version for small systems as the effort required to carry out all the stages, parts and tasks may not be justified. The cost of analysis and design may make the total system unfeasible. Figure 3.10 shows a cut down methodology.

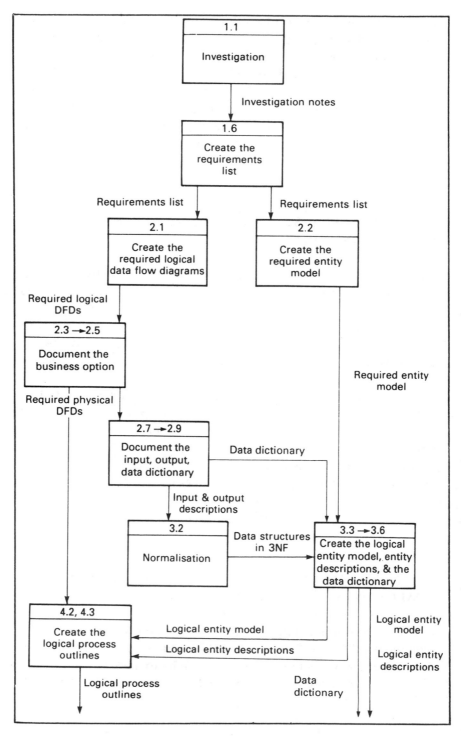

Fig. 3.10

The tasks chosen from part 1 are tasks 1.1, investigation, and 1.6, create the requirements list. The formal modelling of the current system is omitted. The input to tasks 2.1 and 2.2, the creation of the required logical models, is therefore restricted to the requirements list. These tasks now rely much more on the analyst's ability to model the requirements directly, which is feasible with a smaller number of functions and entities. A review of the requirements should now be undertaken with the user before more detailed work commences.

The number of business options with small systems is likely to be very low. Indeed, in many systems, for example those based on micro computers, there may be only one option.

The early cross referencing of data flow diagrams and entity models is omitted. Tasks 3.2 to 3.6 are merged into two tasks to provide a cross reference and proving stage.

The input and output descriptions are converted to third normal form entity descriptions. The entity descriptions are compared with the required entity model to produce the logical entity model and the logical entity descriptions.

Finally, tasks 4.2 and 4.3 are used to create the logical process outlines. Note that the logical process outlines are created without the input of function descriptions or entity life histories. Again this task is feasible for small systems.

The outputs from the logical stages of structured systems analysis and design are identical to the full version. Stage C should now progress as in the full version.

Note that the cut-down version works equally well for totally new systems and replacement of existing ones.

3.7 STRUCTURED SYSTEMS ANALYSIS AND DESIGN FOR NEW DEVELOPMENTS

The development of a new system for a totally new environment is very rare. Few new systems are developed for newly-created organisations or departments. Most systems replace an existing current system.

If no current system exists, then some of the tasks during stage A cannot be carried out. Part 1 would be cut down to tasks 1.1 and 1.6 only, investigation and documentation of requirements.

The part 2 tasks which create the required logical models must now take, as their only input, the requirements list. After this point, the stages and tasks can progress as in the full version.

3.8 THE SYSTEM DEVELOPMENT CYCLE PHASES, STAGES, PARTS, SECTIONS, TASKS AND SUB-TASKS

The systems development cycle comprises seven phases which were described in Section 1.8. Phases 4 and 5, analysis and design, are the subject of this book. Structured systems analysis and design comprises three stages, each stage comprises a number of parts, tasks and sub-tasks. Stage A contains many tasks; for ease of understanding it is divided into sections.
The hierarchy is:

Phases
   Stages
      Parts
         Sections
            Tasks
               Sub-tasks

Figure 3.11 shows all the phases, stages, parts, sections and tasks. It forms a reference for the remainder of the book.

Phase	Stage	Part	Section	Task
1 Terms of reference & business objective				
2 Feasibility study				
3 Investigation	—			
4 Analysis	A Analysis	1 Analysis of the current system	1 Create the current physical model	1.1 Investigate the current system
				1.2 Create the current physical DFDs
				1.3 Create the current entity model
			2 Create the current logical model	1.4 Create the data store entity cross reference
				1.5 Create the current logical DFDs
			3 Create the problems & requirements list	1.6 Create the problems & requirements list
		2 Specification of requirements	1 Create the required logical model	2.1 Create the required logical DFDs
				2.2 Create the required entity model
			2 Create the outline business specification	2.3 Postulate the business options
				2.4 Review and select an option
				2.5 Set the design constraints
			3 Create the detailed business specification	2.6 Create the entity descriptions
				2.7 Prototype and create the I/O descriptions
				2.8 Create the data dictionary
				2.9 Prototype and create the on-line dialogue
				2.10 Create the function descriptions

Fig. 3.11   *continued*

5 Design	B Logical design	4 Validate the specification	2.11 Create the entity function matrix
			2.12 Create the entity life histories
		3 Logical data design	3.1 Select data structures
			3.2 Normalisation
			3.3 Create the entity descriptions
			3.4 Create the 3NF entity model
			3.5 Create the logical entity model
			3.6 Create the logical entity descriptions
		4 Logical process design	4.1 Review the stage A documentation
			4.2 Create the process catalogue
			4.3 Create the logical process outlines
	C Physical design	5 Physical design	5.1 Specify the database
			5.2 Specify the access paths
			5.3 Specify the programs
			5.4 Design tuning
			5.5 Create the implementation plan
			5.6 Create the user manual
			5.7 Create the operations manual
6 Implementation			
7 Operation			

Fig. 3.11   *continued*

PART 2

STAGES, PARTS, AND TECHNIQUES – DETAIL

Chapter 4
Analysis of the Current System

4.1 INTRODUCTION

This chapter describes the first part of structured systems analysis and design – analysis of the current system – shown in Figure 4.1. This part commences with an investigation of the current system and concludes by producing three documents: the current logical data flow diagrams – the user's view of the system; an entity model – a system view; and a combined list of problems with the existing system and requirements for the new system.

The tasks within the first part must be considered as an iterative set of tasks. It is not possible to fully investigate the current system to an extent where no questions will be raised during the creation of the data flow diagrams and the entity model. The creation of the diagrams organises the investigation notes into meaningful models. The execution of these tasks will reveal missing facts and alternative interpretations which can generally only be solved by further investigation.

Task 1.2, creation of the current physical data flow diagrams, and task 1.3, creation of the current entity model, are the tasks which organise the investigation notes into meaningful models. Many analysts experienced with data flow diagrams and entity models use the modelling techniques as their format for investigation notes.

Tasks 1.4 and 1.5 convert the current physical data flow diagrams into the current logical data flow diagrams. The objective is to remove from the data flow diagrams any reference as to how the current system operates. The move is to the logic of the system, what is carried out, not how it is carried out. Task 1.4 creates a set of cross reference diagrams between the physical and the logical to assist the conversion task.

The final task within part 1, task 1.6, produces a combined list of problems with the current system and requirements for the new system. This list will be used in part 2, along with the models of the existing system, to create a specification of requirements. This specification

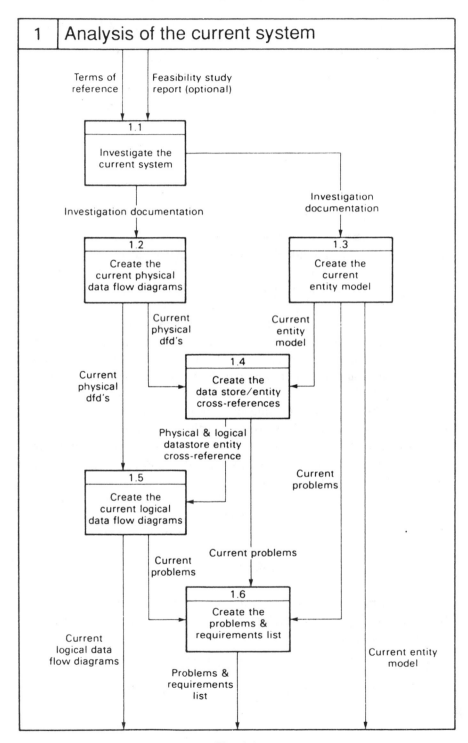

Fig. 4.1

should solve the problems and include the stated requirements for the new system.

Part 1 may be considered as three distinct sections:

Section 1 Creation of the current physical model

Section 2 Creation of the current logical model

Section 3 Creation of the problems and requirements list

4.2 SECTION 1: CREATION OF THE CURRENT PHYSICAL MODEL

4.2.1 Task 1.1: Investigate the current system

The investigation of the current system should include functional analysis, data analysis, problem analysis and requirements for the new system. The objective is to provide sufficient data to be able to construct a set of data flow diagrams, an entity model, and a problems and requirements list.

Investigation techniques, such as the use of questionnaires and interviewing, are skills all systems analysts should possess. These are the techniques required to undertake task 1.1; they are basic techniques required when using any systems analysis and design methodology. The objective of this task is to describe a specific analysis and design methodology assuming the analysts and designers possess the basic skills. Investigation techniques are not, therefore, described in detail. Readers will find a good definition of investigation techniques in Mason and Willcocks (1987), Anderson (1989), and in many texts on systems analysis.

Task 1.1 should provide detailed notes for task 1.2, creation of the current physical data flow diagrams, and task 1.3, creation of the current entity model.

No investigation can ever be 100 per cent complete. Ambiguity as well as missing facts will be revealed during the modelling of the system. Problems can only be sorted out by further discussion and further investigation. For this reason, tasks 1.1, 1.2 and 1.3 should be considered to be an interative loop rather than a fixed sequence of tasks.

Functional analysis

Functional analysis leads to documentation of how the current system operates. The documentation comprises information on the documents within the system, the documents received into the system and those produced by the system, the source and destination of the documents, the storage of documents and the transformation which takes place on documents.

The movement of documents represents data flow, the storage of documents is represented by data stores and the transformation of documents by functions. Functional analysis, therefore, produces a set of data flows, sources of data flow, destinations of data flow, data stores and functions. Investigation, analysis and documentation is carried out top down, high level functions being identified first with lower levels being introduced by successive functional decomposition. Functional analysis provides a good understanding of the system and leads directly to task 1.2, creation of the current physical data flow diagrams.

Data analysis

The objective of data analysis is to identify the logical data within the system. Investigation of the data within the system should be carried out independently of functional analysis. Again the investigation, analysis and documentation should be carried out top down, with the relevant data groups being identified before the content of the data group is considered. Data analysis leads directly into task 1.3, creation of the entity model, where data analysis is discussed in detail.

Problem analysis and requirements

Interviewing is perhaps the most used and most effective way of identifying problems with the current system and requirements for the new one. Many of the problems and requirements will be identified during the investigation for functional analysis and data analysis. These should be carefully noted for documentation later, task 1.6. Additionally, a section of each interview should be reserved for discussion of problems and requirements. Again, these discussions should be carefully noted for later formal documentation.

4.2.2 Task 1.2: Create the current physical data flow diagrams

This task is the first to use data flow diagrams. Before describing current physical data flow diagrams, data flow diagrams (DFDs) in general will be described.

DATA FLOW DIAGRAM

SYSTEM: X-RAY MANAGEMENT	DATE:
AUTHOR: G. CUTTS	PAGE: 1 of 3

LEVEL: 1	CURRENT/~~REQ.~~	PHYS./~~LOGICAL~~

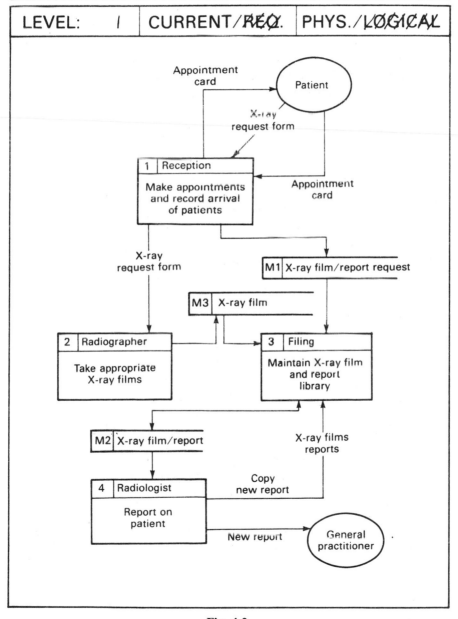

Fig. 4.2

Data flow diagrams

Figure 4.2 shows an example of a data flow diagram (DFD) for a hospital X-ray department, produced from the following investigation notes.

X-ray management investigation notes

Reception   Patients present X-ray request forms, which were obtained from their GP, to reception. Each X-ray request form is allocated an appointment which is written on to an appointment card and given to the patient. X-ray request forms are filed. A diary is maintained which contains details of all the appointments. When patients arrive for X-ray, they present their appointment card. A nurse checks the validity of the appointment, passes the appointment card to a clerk and takes care of the patient. The clerk generates an X-ray film/report request for the filing section. The X-ray request form is retrieved from the file and given to the radiographer. X-ray film/report requests are placed in a temporary file for collection by the filing section.

Radiographer   On receipt of the X-ray request form, the radiographer takes the appropriate photographs (called films) and places them in a temporary file for collection by the filing section. Each appointment results in a set of X-ray films.

Filing clerk   The filing clerks collect the X-ray film/report requests. These request all X-ray films and reports previously generated for the patients. Patients can have many X-ray films and reports. The new X-ray films are attached and the complete set placed in a temporary file for the attention of the radiologist who makes out a report for the appointment. When a copy of the new report is received from the radiologist, all X-ray films and reports are returned to their permanent file.

Radiologist   The radiologist examines all X-ray films and reports which a patient has, and makes out a new report. The report is sent to the GP with a copy to the filing section.

The DFD shows the functions within the system: make appointments and record arrival of patients, take appropriate X-ray films, maintain X-ray film and report library, and report on patients. These functions are carried out currently by reception, radiographers, filing clerks and radiologists respectively. Further, since the total X-ray department function has been decomposed into four functions, then they have been simply numbered one through four. Functions are the dominant feature of data flow diagrams and are represented by a rectangle.

The diagram also shows the sources and destinations for information. These are shown by ellipses, patient and GP. The patient makes an X-ray request, receives an appointment card and returns to the X-ray department with the appointment card. The GP receives a report on the patient. The boundary of the system to be further investigated is also implicitly shown by the diagram. Functions are inside the system and sources and destinations for information outside. The information passing between patient and GP is outside the system and does not cross the boundary; the system can therefore have no effect on this relationship.

Arrows on data flow diagrams illustrate flows of data. In the example, the flows of data represent documents and X-ray films. For example, an X-ray request form flows from reception to the radiographers and a copy new report flows from the radiologists to the filing clerks. These data flows are internal flows as opposed to the data flows appointment card, X-ray request forms and new report which form the input and output of the system.

Finally, the data flow diagram shows open-ended rectangles to indicate stores of data. The description data store is carefully chosen since data is stored in many forms including files. For example, the data stores in the X-ray department could be an in-tray of requests for the filing clerk to retrieve previous X-rays and reports from a filing cabinet, and a simple pile of previous X-rays and reports together with the new X-ray films for reference by the radiologist. Further data stores will be contained within the functions.

Data flow diagrams provide a user's view of the system showing:
• Functions by rectangles
• The sources of data by ellipses
• The destinations of data by ellipses
• Data flows by arrows
• Data stores by open-ended rectangles

Templates and sophisticated software packages such as Automate Plus, Excellerator and Systems Engineer are available for the construction of data flow diagrams. These packages do much more than diagramming, they are well developed CASE tools.

DATA FLOW DIAGRAM

SYSTEM: X - RAY MANAGEMENT   DATE:

AUTHOR: G. CUTTS   PAGE: 2 of 3

LEVEL: 2   CURRENT/~~REQ~~.   PHYS./~~LOGICAL~~

TITLE: RECEPTION

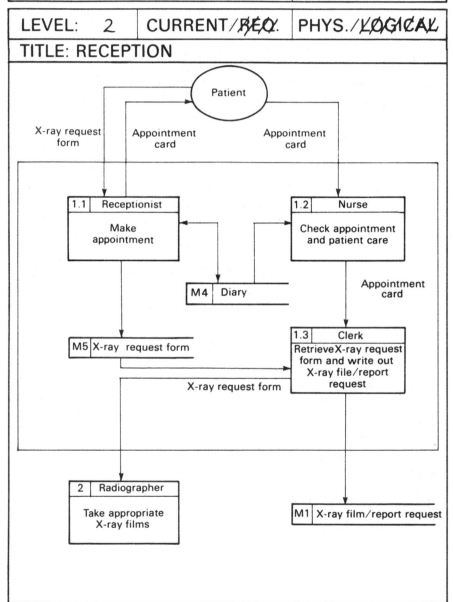

Fig. 4.3

Levels of decomposition

The X-ray department shown in Figure 4.2 has been decomposed into four functions. There are no rules governing the number of functions that should be shown on a single DFD other than that of understandability. This means that any more than, say, ten functions on a single diagram may lead to confusion. The basic concept of the DFD is that it represents the user's view and must therefore be totally understandable by the user.

Further decomposition, and thus a greater level of detail and understanding, is obtained by zooming in on and decomposing any or all of the function rectangles. Figure 4.3 shows a further decomposition of the 'make appointments and record arrival of patients' function. Note that the input and output exactly correspond with the data flows corresponding to function 1, Figure 4.2.

Note also that since the functions of make appointment, check appointment and patient care, and retrieve X-ray request and write out X-ray film/report request, are all sub-functions of function 1, they are numbered 1.1, 1.2 and 1.3 respectively. It is also possible to conclude that the receptionists, nurses and clerks are all part of the staff of reception. Two new stores of data also emerge within reception, a diary and a file of X-ray request forms.

Figure 4.2 is labelled level 1 DFD, which may be decomposed into up to four level 2 DFDs, which in turn may be further decomposed to levels 3 and 4. Rarely are more than four levels of decomposition necessary, even for the largest most complex systems.

Decomposition should cease when the function can be described accurately, precisely and unambiguously. This may well mean that some functions have more levels of decomposition than others. Each function at its lowest level, its primitive level, should process only a single transaction type. For example, the transaction types for a sales accounting system might be: issue an invoice, issue a credit note and receive a payment. A primitive function should exist for each type of transaction.

Figure 4.4 shows the level 2 DFD for the level 1 function, maintain X-ray film and report library.

DATA FLOW DIAGRAM

SYSTEM: X-RAY MANAGEMENT	DATE:
AUTHOR: G. CUTTS	PAGE: 3 of 3

LEVEL: 2	CURRENT/~~REQ.~~	PHYS./~~LOGICAL~~

TITLE: FILING

| M1 | X-ray film/report request |

| 3.1 | Filing clerk |
Retrieve all existing X-ray films & reports for this patient

| M8 | X-ray film |

| M6 | Report |

| M7 | X-ray film/report |

| 3.2 | Filing clerk |
Merge new film with existing in date sequence

| 3.3 | Filing clerk |
Insert films & reports back in files

X-ray films/reports
Copy new report

| M3 | X-ray film |

| M2 | X-ray film/report |

| 4 | Radiologist |
Report on patient

Fig. 4.4

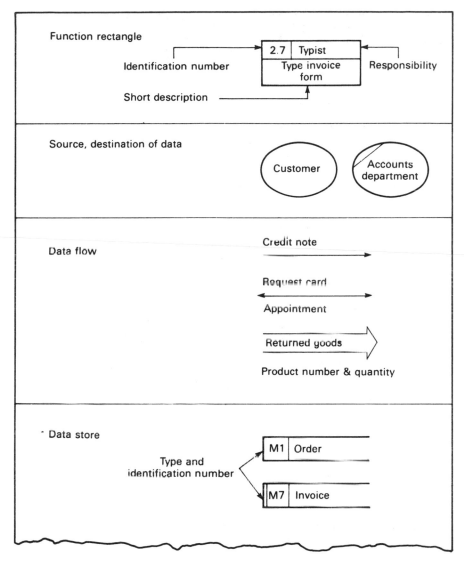

Function rectangle

Identification number

Short description

Source, destination of data

Data flow

Data store

Fig. 4.5

Diagram conventions (Figure 4.5)

Functions are represented by rectangles containing a short but meaningful description. The function is given an identification number – a single number at level 1 with compound numbers at subsequent levels

– and are annotated with the responsible authority for completing the function.

The description must contain a verb that is as descriptive as possible. The use of verbs such as 'update' and 'process' should be avoided as they are not highly descriptive.

The responsibility box will commence as the person or department responsible and may conclude as a program reference. If the function remains external to the computer system, the new person or department responsible may be inserted into the responsibility box.

The ellipse is used to show the source or destination of a data flow. Each ellipse should be carefully named. To avoid over-complication, the ellipse may be repeated on a DFD. To show that there exists more than one ellipse representing an external source or destination, a line is introduced into the top left hand corner of the ellipse. This line is then present in every occurrence of the ellipse.

Data flows are shown by arrows suitably annotated. The arrow may be double-headed to show, for example, the reading and subsequent modification of a data store. Occasionally, the movement of goods represents a data flow. It is, however, more frequent that goods movement does not represent a flow of data and, therefore, should not be shown on a DFD. In Figure 4.5, returned goods represent a data flow made up of product number and quantity of goods returned. This data is processed within the system. The movement of goods is represented by a broad arrow.

The final diagram structure is the open-ended rectangle to represent a data store. The name data store is carefully selected to represent not only files but also wall charts, lists, private reference books etc., in fact any source of data within the system. Data stores are named, e.g. invoices, given an identification number and a type. Types are typically M for manual data store and C for computer file. Finally, the double bar at the closed end of the rectangle indicates that this data store is repeated elsewhere on the DFD.

Data flows into or out of data stores need not be named as part of this task.

Figure 4.5 shows a function 'type invoice form' which is a decomposed function from function 2. It is the responsibility of a typist. Figure 4.5 further shows a customer source or destination and an occurrence of the accounts department as a source or destination. Data flows shown are credit note and a two way flow regarding appointments. Also, the flow of returned goods represents the data flow product number and quantity. Finally, Figure 4.5 shows an occurrence of the manual data store, invoices, identification number seven. It is only one of the occurrences on the diagram since the open-ended rectangle possesses a double line at its closed end.

Developing the current physical data flow diagram

The first use of DFDs occurs in part 1, task 1.2, creation of the current physical data flow diagrams. The diagrams are current in that they reflect the system as it exists, the current system, and they are physical in that they reflect how the system operates. The current physical DFDs model functions which, at level 1, might be the departments or sections; the sources – providers of data of the system; and the destinations – users of data from the system. Document flows and data stores, representing current files, lists, wall charts, reference books etc., are also modelled on the current physical DFDs.

Three sub-tasks make up task 1.2:

1.2.1   Development of a document flow diagram.
1.2.2   Conversion of the document flow diagram to the current physical level 1 data flow diagram and decomposition to subsequent levels.
1.2.3   Validation of the diagrams.

Sub-task 1.2.1: Document flow diagrams

This sub-task identifies the source and destination of data and its flow between them. The only diagram structures required are the ellipse and the arrow.

Data flows represented by arrows are real document or goods flows. Sources and destinations of data may be external to the organisation, e.g. accounts department, senior management or existing computer systems such as the sales ledger.

The interview notes should be read and a list of sources and destinations created with a second list of documents. The document flow diagram can now be drawn. Figure 4.6 shows a simple document flow for the sources, destinations and document flows listed below.

Sources and destinations
   Customer
   Despatch
   Warehouse

DOCUMENT FLOW

SYSTEM: *EXAMPLE*	DATE:
AUTHOR: *G CUTTS*	PAGE: I of I

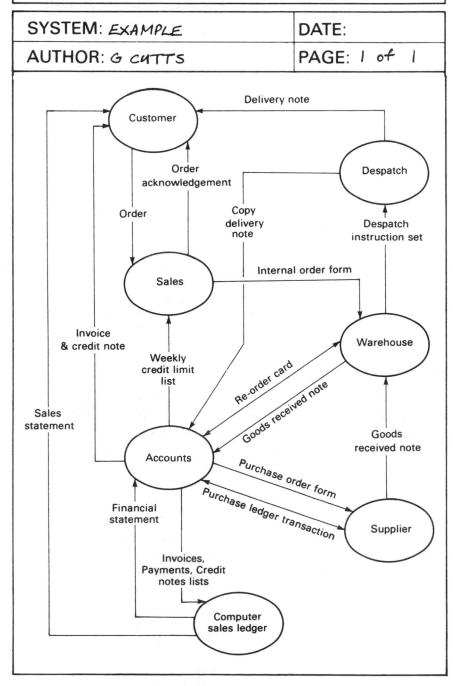

Fig. 4.6 Document flow

Sales
Accounts
Supplier
Computer sales ledger

Documents
Delivery note and copy
Despatch instruction set
Order
Internal order form
Order acknowledgement
Weekly credit limit list
Re-order card
Goods received note
Purchase order form
Purchase ledger transactions
Invoices, payments and credit notes list
Invoice and credit note
Financial statement
Sales statement

The diagram can be built in real time. That is, as a direct result of an interview. Many experienced systems analysts find it easier to draw document flow diagrams than to take interview notes.

The document flow diagram should be shown to the user and agreement on its accuracy obtained.

The final step in sub-task 1.2.1 is to agree the boundary of the system with the user management. Figure 4.7 shows a system boundary. The required system to be further investigated and analysed comprises functions for sales, despatch and the warehouse with sources or destination of data, customer, accounts and supplier. The data flows from accounts to supplier and accounts to computer sales ledger are completely outside the system; they do not cross the system boundary nor are they within the boundary. The new system, therefore, can have zero effect on these data flows. Several boundaries may be discussed before agreement is reached.

Sub-task 1.2.2: Data flow diagrams

The level 1 current physical data flow diagram shown in Figure 4.8 is now easily constructed by redrawing the document flow shown in Figure

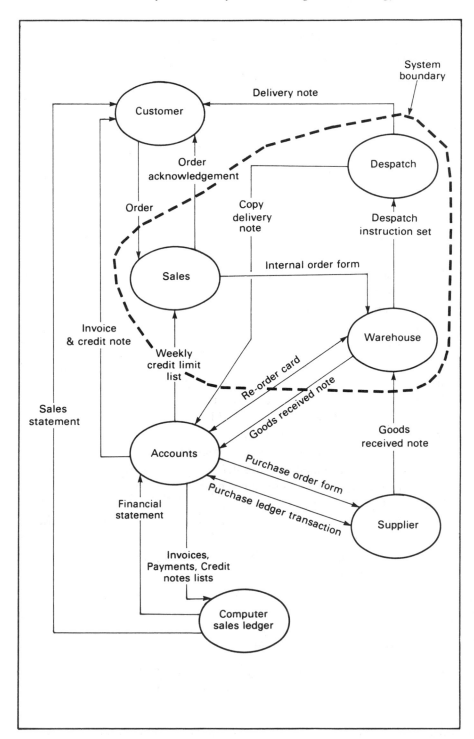

Fig. 4.7

DATA FLOW DIAGRAM

SYSTEM: CHAPTER 4 EXAMPLE	DATE:
AUTHOR: G. CUTTS	PAGE: 1 of 3

LEVEL: 1	CURRENT/~~REQ.~~	PHYS./~~LOGICAL~~

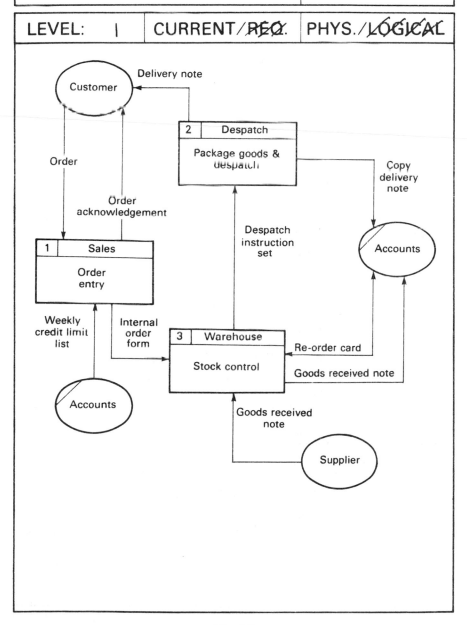

Fig. 4.8

4.7, replacing sales, despatch and warehouse by function rectangles and by omitting all sources, destinations and data flows not affecting the system.

Level 2 current physical data flow diagrams can then be drawn for each of the level 1 functions. A simple example is shown in Figures 4.9 and 4.10: the decomposition of 3, stock control, and 3.3, monitor stocks levels.

Decomposition

The diagram must be agreed with the user before decomposition of the level 1 current physical data flow diagram is commenced. A structured walkthrough is the best approach. The user can become totally involved with DFDs and often give a high level of commitment to the accuracy of the diagrams.

Two potential problems arise at the point of decomposition. They are the identification of sub-functions and the incorporation of data stores.

The investigation notes identified the level 1 functions as data sources or destinations. These level 1 functions may represent departments within an organisation such as accounts. The sections within the department may form the level 2 functions. These sections may be identifiable from the existing investigation notes or it may be necessary to return to task 1.1, investigation, to obtain more information about the sections within the department. The first investigation would have regarded the department's inputs and outputs of major importance to the production of the document flow diagram. Since the department is inside the boundary, a further investigation may be necessary to determine the internal data flows, data stores and level 2 functions.

The boundary of the level 2 DFD is easily obtained; it is the function rectangle on the level 1 diagram. Similarly, the input, output, sources and destinations are defined by the level 1 diagram (see Figure 4.8). Figure 4.11 shows the boundary of function 3, stock control.

Each input or output must connect to a function. The easiest starting point for construction of the level 2 diagram is by identification of the functions which receive the inputs. The functions can be added to the diagram, Figure 4.11, to give Figure 4.12.

The functions which generate output should now be examined. Ask the question, 'Does the process have access to all the data necessary to generate the output?' The answer will most likely be 'No'. The inputs to

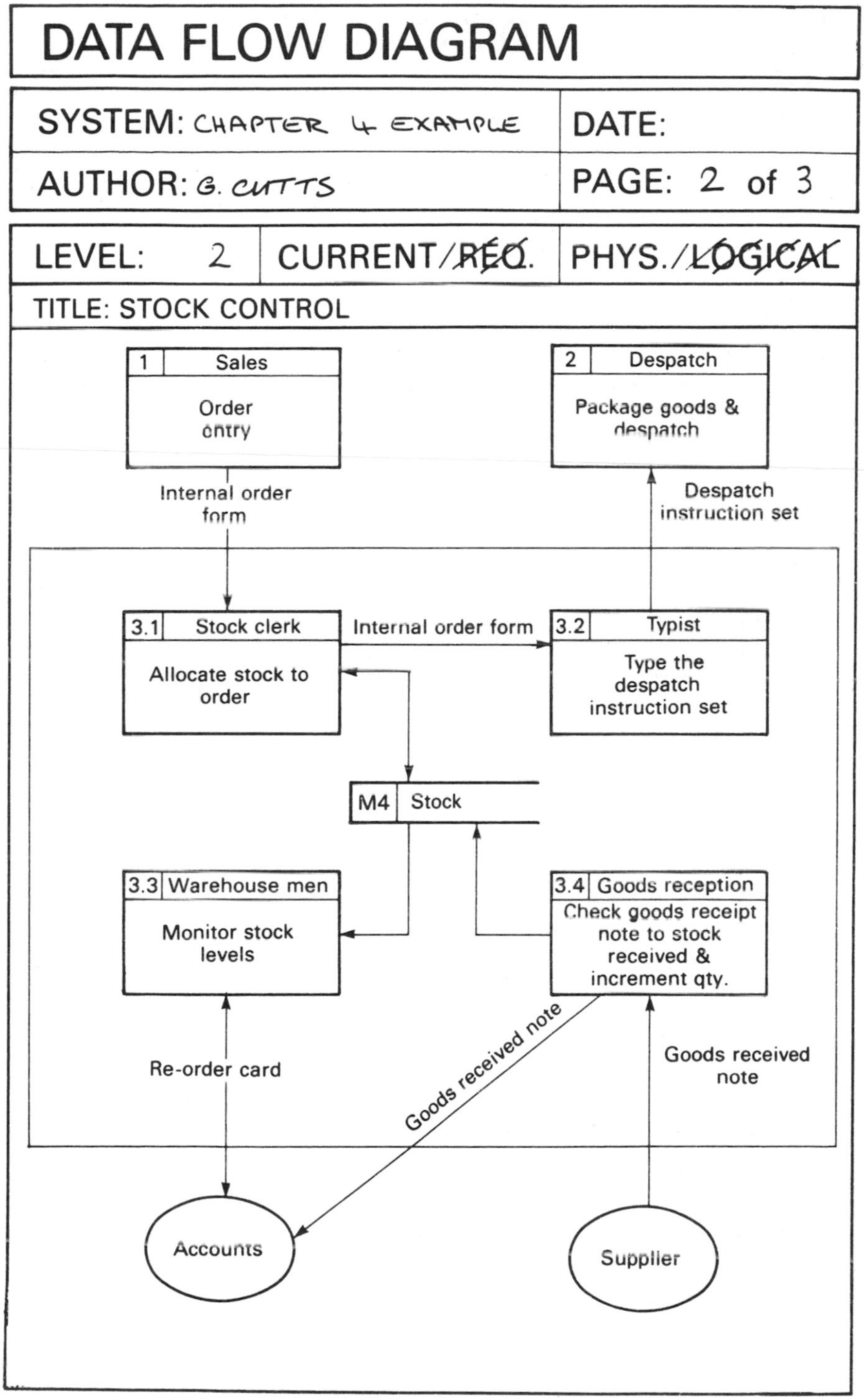

Fig. 4.9 Stock control

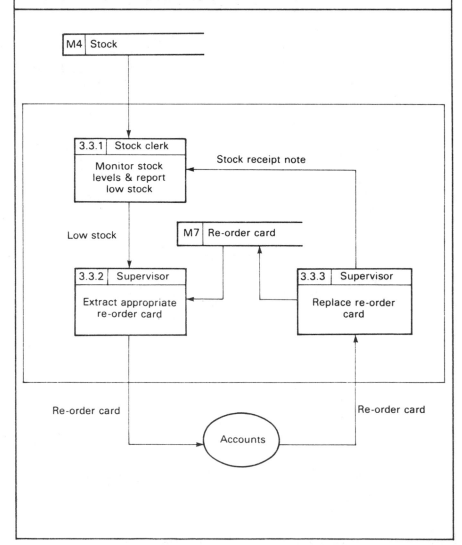

DATA FLOW DIAGRAM

SYSTEM: CHAPTER 4 EXAMPLE	DATE:
AUTHOR: G. CUTTS	PAGE: 3 of 3

LEVEL: 3	CURRENT/~~REQ.~~	PHYS./~~LOGICAL~~

TITLE: MONITOR STOCK LEVELS

M4 Stock

3.3.1 Stock clerk
Monitor stock levels & report low stock

Stock receipt note

Low stock

M7 Re-order card

3.3.2 Supervisor
Extract appropriate re-order card

3.3.3 Supervisor
Replace re-order card

Re-order card

Re-order card

Accounts

Fig. 4.10 Monitor stock levels

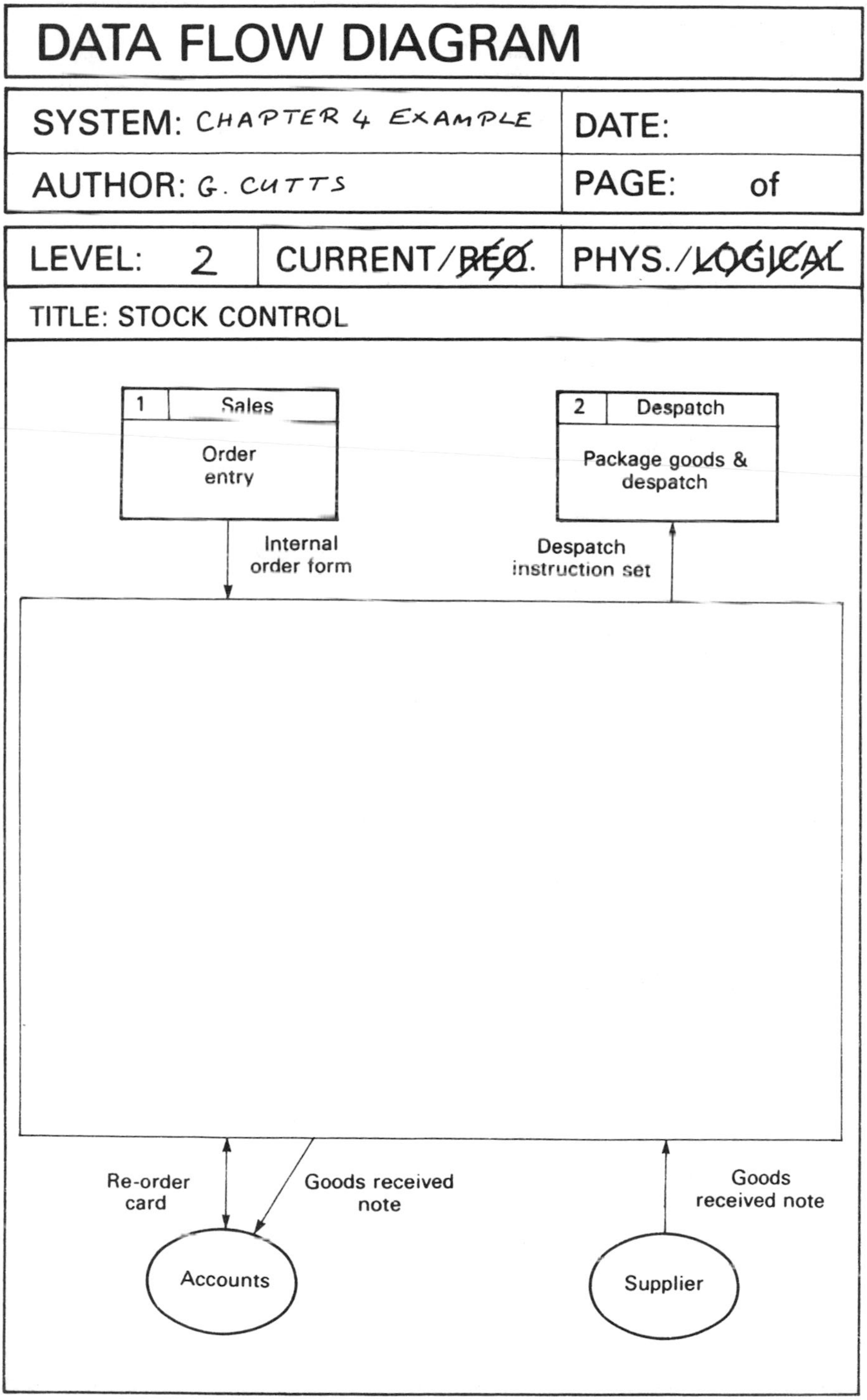

┌───┐
│ # DATA FLOW DIAGRAM   │
└───┘

SYSTEM: CHAPTER 4 EXAMPLE   **DATE:**

AUTHOR: G. CUTTS   **PAGE:** of

LEVEL: 2   **CURRENT/~~REQ~~.**   **PHYS./~~LOGICAL~~**

TITLE: STOCK CONTROL

1	Sales
	Order entry

2	Despatch
	Package goods & despatch

Internal order form

Despatch instruction set

Re-order card

Goods received note

Goods received note

(Accounts)

(Supplier)

Fig. 4.11 Stock control – level 2

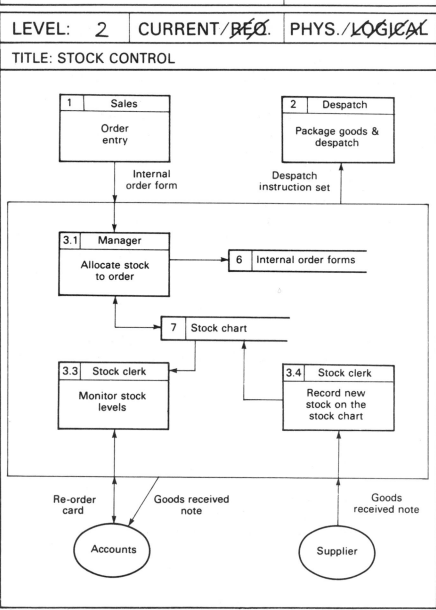

DATA FLOW DIAGRAM

SYSTEM: CHAPTER 4 EXAMPLE   DATE:

AUTHOR: G. CUTTS   PAGE: of

LEVEL: 2   CURRENT/~~REQ.~~   PHYS./~~LOGICAL~~

TITLE: STOCK CONTROL

1	Sales
	Order entry

2	Despatch
	Package goods & despatch

Internal order form

Despatch instruction set

3.1	Manager
	Allocate stock to order

6	Internal order forms

7	Stock chart

3.3	Stock clerk
	Monitor stock levels

3.4	Stock clerk
	Record new stock on the stock chart

Re-order card

Goods received note

Goods received note

Accounts

Supplier

Fig. 4.12 Stock control – level 2

the generation process must be investigated. They will either originate from data stores or from further internal functions. In some instances, an internal function will connect inputs to outputs. The answer to the question, 'Does the process have access to all the data necessary to generate output?' may in these cases be 'Yes'. Manual functions such as 'write out the internal order form' may connect the input 'customer order' to the output 'internal order form'.

Data stores and functions can now be added until the diagram is complete. It is worth emphasising again, at this stage, that the DFDs represent the current physical system. Data stores represent wall charts and private books of data as well as current files; functions have their responsibility annotated and data flows are very often document flows.

The problem of when to introduce a data store on to the diagram always needs careful handling. Below are three instances of current systems which might yield the diagrams shown in Figure 4.13.

(1) Documents are date stamped and checked before being left in the out-tray for collection by order processing.
(2) Documents are date stamped and checked before being placed in the central store room for order processing to collect.
(3) Documents are date stamped and checked before being sent directly to order processing where they are stored prior to processing.

Each diagram is correct. The data store is always present and is introduced on to the diagrams at an appropriate point. Data stores which are completely internal to a function need not be shown at that level, which is the case in situations (1) and (3).

In situation (1), documents are stored within function 1.1 and a batch of documents flows to the sales department; whereas in situation (3), documents flow individual to be stored within function 1.2.

Data flows to or from data stores need not be named at this moment.

Sub-task 1.2.3: Validation of the diagrams

The first step in the validation of the diagrams is to check for inconsistencies.

Functions should not be total sources of data or destinations for data. Those which are total sources generate data without reference to other data. Those which are total destinations perform no useful purpose. Generally, if such a situation exists, then further investigation is necessary to establish the exact processing performed.

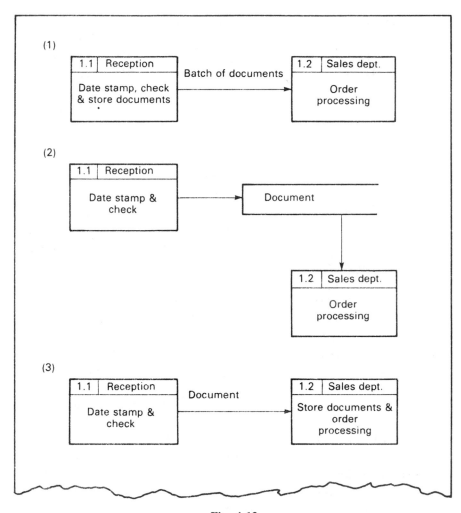

Fig. 4.13

Similarly, data stores should not be total providers of data nor total destinations for data. Data is stored by one function to be used by another. There are, however, exceptions. The data store 'archived pay advice' is necessary for legal purposes. Pay advices are required to be stored for a set period just in case a query may arise in later years. This query function may not be shown on the DFD since it is of extremely low volume and includes a wide variety of query processing. The data store 'archived pay advices' may therefore be a legitimate total destination.

There are also situations where a file, particularly a computer file, is maintained by one system and used solely as a reference file by another. The latter system would show the file as a total source. All situations where data stores are total sources or destinations should be investigated and reasons established for the situation, or additional data flows created if necessary.

Finally, the validation process should examine data flows. All data flows should have a source, an external entity, a function or a data store, and a destination, again an external entity, a function or data store. Inevitably, some data flows will not have a destination. These again require further investigation to determine their destination or, if none can be identified, their need to be generated in the first place. Similarly, some data flows will have no source; again this is a situation for further investigation and clarification.

The context diagram

Figures 4.7 and 4.8 effectively show the boundary of the system. If there is any problem in clearly understanding the boundary, a level zero current physical data flow diagram may be drawn. This diagram puts the complete system into context and is sometimes termed the context diagram. Figure 4.14 shows a level zero current physical data flow diagram for a system comprising the functions of sales, warehousing and despatch.

The system is represented as a single function, making absolutely clear that everything outside the function is external to the system. The input and output data flows and their sources or destinations are equally clear. The context or level zero data flow diagram is a useful representation of the system boundary, together with the system input and output.

The task of constructing the current physical data flow diagrams has so far followed a series of sub-tasks in sequence. There are, however, many instances during the task when iteration to task 1.1, investigation, is necessary. This may be to sort out a problem discovered during validation. Other reasons for re-opening the investigation arise from problems in agreeing the boundary and in identifying the functions and data stores during decomposition.

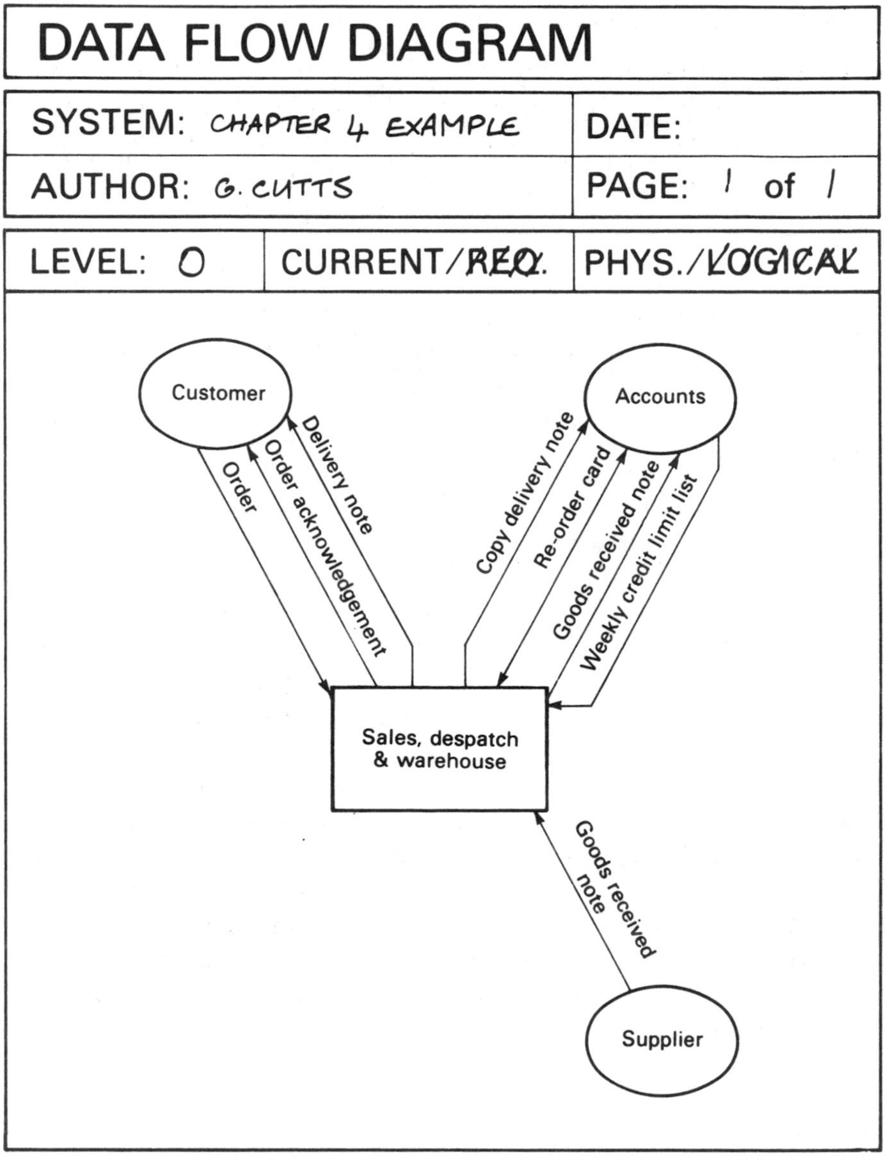

DATA FLOW DIAGRAM

SYSTEM: CHAPTER 4 EXAMPLE	DATE:
AUTHOR: G. CUTTS	PAGE: 1 of 1

LEVEL: O	CURRENT/~~REQ~~.	PHYS./~~LOGICAL~~

Fig. 4.14

Problems with boundary identification

The boundary shown in Figure 4.7 neatly encloses all of the functions undertaken by sales, the warehouse and despatch. If only part of the functions for the warehouse were required to be inside the boundary, then a diagram similar to Figure 4.15 would result.

The warehouse function needs to be decomposed to identify those sub-functions which fall inside the boundary and those which fall outside. The inside sub-functions and the outside sub-functions can be grouped to form two functions representing the warehouse. In this case, shown in Figure 4.16, the inside function is concerned with goods out and the outside function with goods in. The system boundary can now be drawn, and the construction of the current physical data flow diagrams can proceed.

4.2.3 Task 1.3: Create the current entity model

This task is again a first in that it introduces entity models which will be used in all stages. Entity models result from data analysis, whereas the DFDs resulted from functional analysis. Data analysis provides a way of structuring data, removing inconsistencies and testing the use of data before detailed design.

Data analysis must be undertaken independently of functions or processing to enable data to be shared. The analyst must take a wide view of the data, a top down approach, producing a view of the data and the relationships between data, inside the system. It is an internal view within the boundary of the system of the data and of the data relationships, and represents the generic underlying structure of the data. This view, a system view, contrasts with the use of DFDs which produce a user's view.

Definitions

Each entity represents a data group of interest to the system. A sales and accounting system may have data grouped together relating to customers, orders, invoices and products. The entities would, therefore, be customer, order, invoice and product.

Each entity represents a data group. A group of data items, called attributes, are held in the system for each entity. Examples of attribute

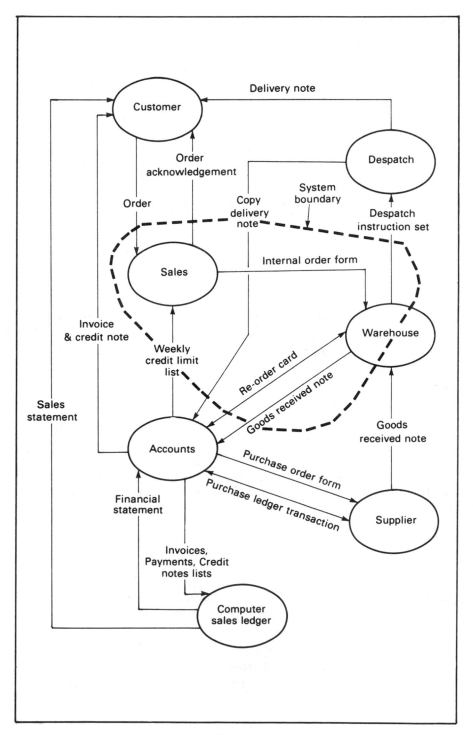

Fig. 4.15

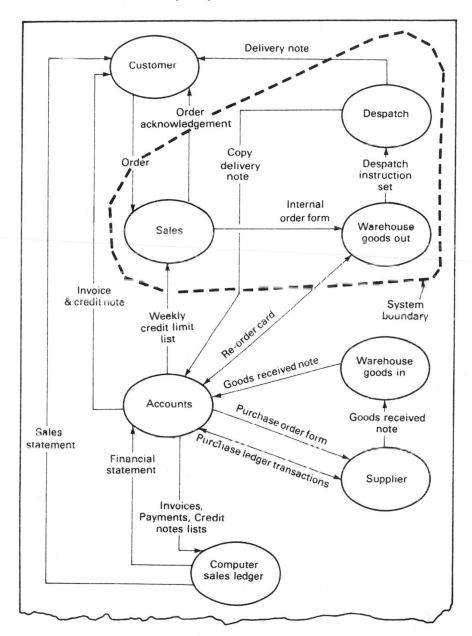

Fig. 4.16

lists for each of the entities – customer, order, invoice and product – are given below:

Customer
    Customer number (key)
    Customer name
    Customer address
    Credit limit
    Delivery address
    Date of last order

Order
    Customer number (key)
    Product number (key)
    Quantity ordered

Invoice
    Invoice number (key)
    Invoice date
    Customer number
    Product number
    Quantity delivered
    Total value

Product
    Product number (key)
    Product description
    Price each

Entities can be identified by extracting the nouns or noun phrases from the investigation notes. These nouns or noun phrases provide a list of candidate entities.

The X-ray management investigation notes might yield the following set of nouns as candidates for entities:

Patient   X-ray request
Diary   Appointment
Report   X-ray film/report request
X-ray film

It is better at this stage to identify too many possible entities than too few. The entity model will be revised later during stage B.

A matrix, Figure 4.17, can then be constructed. In fact, only half is

	Patient	Diary	X-ray request	Appointment	X-ray/report request	X-ray film	Report
Patient							
Diary							
X-ray request	Presents						
Appointment		Contains	Is allocated				
X-ray film/ report request				Generates			
X-ray film	Has			Results in			
Report	Has			Results in			

Fig. 4.17

required since the matrix will be used to indicate relationships between entities.

A second analysis of the investigation notes, this time for verbs, provides candidate relationships between the entities. For example, 'a patients present X-ray request forms to reception':

Entity   Patient
Relationship   Present
Entity   X-ray request

or 'a patient has many X-ray films and reports':

Entity   Patient   Patient
Relationship   Has   Has
Entity   X-ray film   Report

The relationships can be marked on to the matrix (Figure 4.17).

A good method of double checking is to compare the entities in pairs. For each pair ask the question, 'Does a relationship exist between this pair of entities?'

Relationship degree

The following entities and relationship have degree 1 : 1 (read 1 to 1):

Entity   Husband
Relationship   Has
Entity   Wife

A husband can only have one wife and a wife can only have one husband (well, to be legal in the UK). The relationship is 1 : 1.

Most relationships have degree 1 : M (read 1 to many):

Entity   Patient
Relationship   Has undergone
Entity   Operation

A patient can have many operations but an operation can have only one patient (transplants are regarded as two operations). The relationship is 1 : M.

Some relationships are M : N (read many to many) where M is not necessarily equal to N:

Entity   Ship
Relationship   Docks
Entity   Port

A ship docks at many ports and a port has many ships. The relationship is M : N where the number of ports per ship is usually different from the number of ships per port.

Diagram conventions (Figure 4.18)

Entities are represented by rectangles annotated with the entity name, which should be meaningful. The attribute which provides the key to the entity may provide a meaningful name.

Lines are used to represent relationships, with the crow's foot representing the many parts of the relationship. The one entity of the relationship is called the owner and the many entity is called the member. Owner entities have many member entities. The crow's foot is sometimes replaced by an arrowhead.

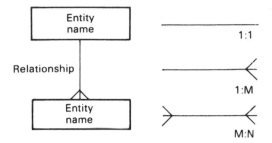

Fig. 4.18

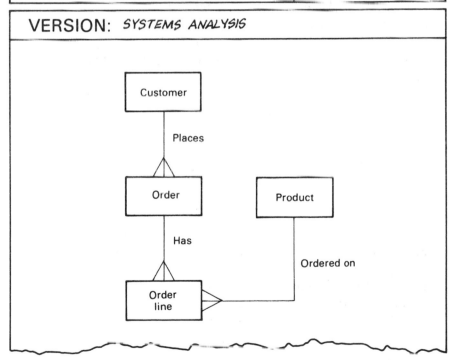

Fig. 4.19

Figure 4.19 shows several entities and relationships. The entities are customer, order, order line and product. The three relationships are shown below:

Entity	Relationship	Degree	Entity
Customer	Places	1 : M	Order
Order	Has	1 : M	Order line
Product	Ordered on	1 : M	Order line

A customer places many orders, each order has many order lines, each order line being for one product. An order therefore can be for one or more products. A product has many order lines from different orders; each order line is from one order and each order is from one customer.

Note that customer is the owner of the customer-order relationship and order is the member; order is the owner of the order-order line relationship and order line is the member; product is the owner of the product-order line relationship and order line is the member. The order entity is both a member and an owner and the order line entity is a member of two relationships.

An entity model represents the complete set of entities and relationships for a system. The version of the entity model created at this stage represents the current understanding at this early stage within the project.

Figure 4.20 shows the entity model for the X-ray system drawn directly from Figure 4.17.

Many: many relationships

The entity models drawn so far have only used 1 : 1 and 1 : M relationships. Many to many relationships create a problem concerning the storage of some data items. Figure 4.21 shows a M : N relationship. A patient may consult many doctors and a doctor will certainly deal with many patients. A list of the attributes for patient and doctor shows some common data items; the problem is where should these common data items be stored.

ENTITY MODEL

SYSTEM: X-RAY MANAGEMENT	DATE:
AUTHOR: G. CUTTS	PAGE: 1 of 1

VERSION: SYSTEMS ANALYSIS

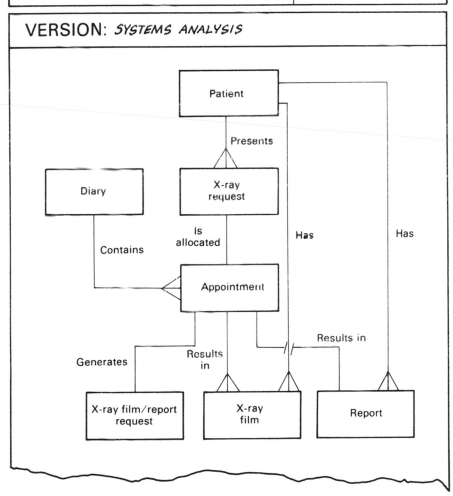

Fig. 4.20

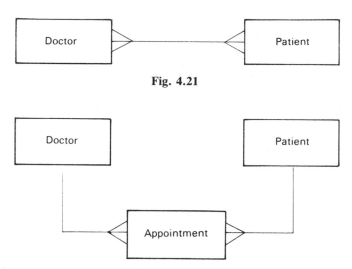

Fig. 4.21

Fig. 4.22

Patient:	Patient number	*Doctor*:	Doctor identification
	Name		Name
	Address		Qualifications
	Health history		Specialism
	List of appointments		List of appointments
	date		date
	time		time
	location		location

The patient has a list of appointments, with several doctors; the doctor has a list of appointments, with several patients. A single appointment is common to one patient and one doctor. The relationship patient to appointment is 1:M, and the relationship doctor to appointment is 1:M; appointment is common to both entities. The M:N relationship shown in Figure 4.21 can be replaced by two relationships and the creation of a third entity, appointment, representing the common data from the two original entities. The entity model which results is shown in Figure 4.22.

Note that order line in Figure 4.19 provides a link between order and product. An order is for many products and a product is for many orders.

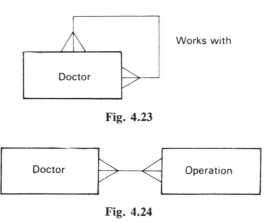

Fig. 4.23

Fig. 4.24

One: one relationships

One to one relationships create some problems which need to be resolved. The first is which entity to make the owner and which the member. One way round the problem is to merge the entities into one. Consider two entities, enquiry and quotation. If one enquiry always results in one quotation, the relationship is 1:1 and one entity will suffice, enquiry/quotation, i.e. the entities may be merged.

A second solution is to examine the time dependencies. For example, consider an order line entity which will have one invoice line entity; the order line entity will exist in the system from order entry through stock allocation processing and delivery before the invoice line entity is inserted. In this case the order line should be the owner.

In many instances, neither of the above will apply. In this case the choice of owner is arbitrary.

Convoluted relationships

The entity model shown in Figure 4.23 demonstrates a new type of relationship. The entity doctor appears on both ends of the relationship: doctor works with doctor. The relationship, works with, requires further investigation. It may mean doctor works with, on an operation, doctor. The common data is the operation; the entity operation can therefore be inserted to link doctor to doctor shown in Figure 4.24. This results in a M:N relationship. The common data between doctor and operation is required for a link. A possible link is the tasks performed, such as surgeon or anesthetist shown in Figure 4.25.

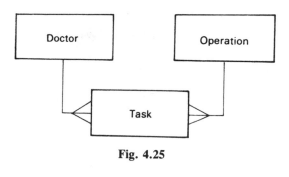

Fig. 4.25

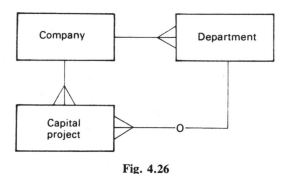

Fig. 4.26

Optional relationships

The entity model shown in Figure 4.26 demonstrates three entities: company, capital project and department. The relationships between company and project and company and department are clear. A company is organised into departments and undertakes capital projects. However, some projects are solely for a particular department; in this case there is a link between department and project. Projects are always undertaken for the company and optionally for a department.

The optional link is shown by the standard relationship arrow annotated with the letter O for optional. These optional relationships will be treated differently in stage C, physical design.

Multiple relationships

All of the entity models constructed so far have shown a maximum of one relationship between an owner and a member. Figure 4.27 shows an entity model where two relationships connect order and order line.

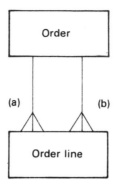

Fig. 4.27

Relationship (a) represents orders with outstanding order lines, relationship (b) represents orders with despatched order lines. This is quite common and basically divides the order lines into two groups: order lines awaiting despatch, and order lines that have been despatched.

Relationship exclusivity

There are instances in the definition of an entity model where exclusivity is important. Consider a situation where an invoice may have a payment or a credit, but not both. We use the relationship icon with a round bracket to show this exclusivity, shown in Figure 4.28.

Exclusivity can also be important from member to owner. Figure 4.29

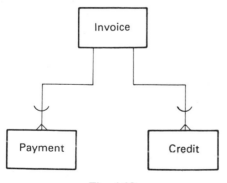

Fig. 4.28

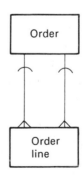

Fig. 4.29

shows an order line which is either outstanding or despatched. It is an enhancement of Figure 4.27.

4.3 SECTION 2: CREATION OF THE CURRENT LOGICAL MODEL

4.3.1 Task 1.4: Create the data store entity cross references

The physical data store entity cross reference

The inputs to task 1.4 are the data stores from the current physical data flow diagrams and the entity model. Every data store on the current physical data flow diagrams must be represented by one or more entities on the entity model. If a cross reference cannot be found, then either a data store has been inserted or, more likely, the entity model is not complete. This type of cross referencing and high level proving during the early stages of the project is a feature of the methodology which leads to good quality systems being produced.

Figure 4.30 shows two data stores on a DFD and an entity model. Customers place orders, where each order contains many order lines, a single order line for each product ordered. For each despatch against an order line, normally only one but possibly more than one, an advice note is sent with the goods. A copy advice note is used by a typist, function 1.7, to create an invoice for each despatch. Function 1.7 references a name and address list to check the invoice address. Copy invoices are sent to accounts where a clerk inserts invoice data on to the sales ledger, function 2.3. An order line has an invoice for each despatch and, therefore, may have many invoices. Payments against invoices or the part credit of invoices are also processed by the accounts clerk. Figure 4.31 shows the cross reference. Note that the full customer entity appears in both existing data stores. This shows a duplication of this data in the current physical system. Note, also, that the invoice entity contains a reference to the order line entity; the order line entity is not present in the sales ledger data store. This is shown using the relationship without the owner entity box.

Figure 4.32 shows the cross reference for the X-ray system. Note that patient data is repeated in manual data stores 1, 2, 4 and 6, and that patient data together with X-ray films and patient reports can be physically located in one of three manual files: 2, 7 and 8.

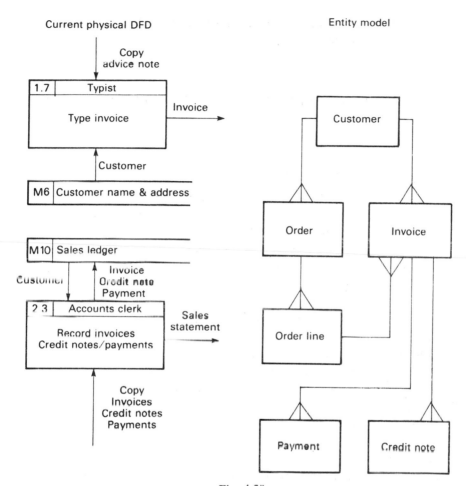

Fig. 4.30

Many of the current physical data stores contain data on more than one entity. This leads to storage anomalies which will be discussed in Chapter 7. Finally, note that data store M3 only includes a reference to the patient entity, name and date of birth perhaps, whereas data store M6 contains the full patient data. This is because of the problems of including any more than a minimum amount of text on an X-ray film.

The current physical data stores may contain several entities. Functions which access data stores may not require access to all of the entities, and it is therefore good practice at this time to return to the

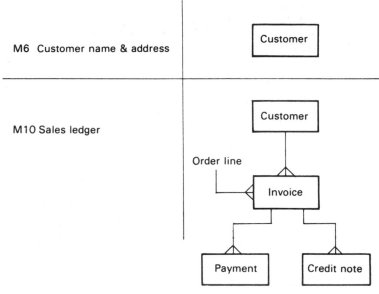

Fig. 4.31

current physical data flow diagrams and to annotate each of the data flows to or from data stores with the names of the entities accessed. These names will be useful during task 1.5, the creation of the current logical data flow diagrams.

Figure 4.33 shows a section of a current physical data flow diagram with data flows to and from the data store sales ledger annotated. Figure 4.34 shows an example of a sales statement with the data from each of the four entities customer, invoice, payment, and sales statement heading.

To complete the cross reference, it is necessary to investigate the attributes within each physical data store and to group them according to the entity names. Examine the entry from the sales ledger shown in Figure 4.34: it contains customer data, invoice data, payment data, and data which can only be allocated to the sales statement itself. This exercise provides a first look at the entity descriptions, i.e. the attributes which make up the entity. Entity descriptions will be formally completed during part 2 but a start can be made now by recording candidate attributes for each entity.

DATA STORE/ENTITY X REF.

SYSTEM: X-RAY MANAGEMENT	DATE:
AUTHOR: G. CUTTS	PAGE: 1 of 1

PHYSICAL/~~LOGICAL~~

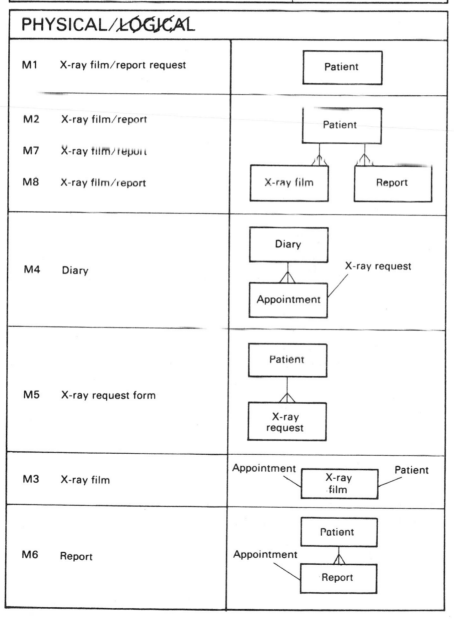

Fig. 4.32

Data store entity cross-reference

M7   Sales ledger

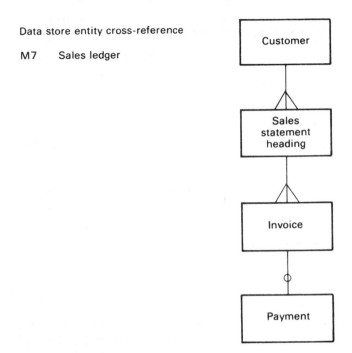

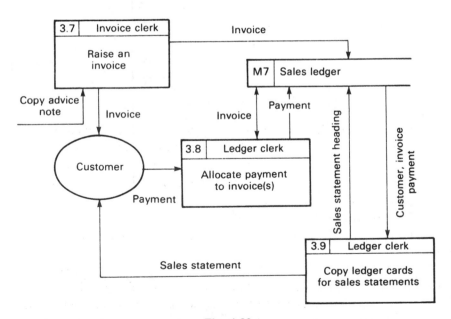

Fig. 4.33

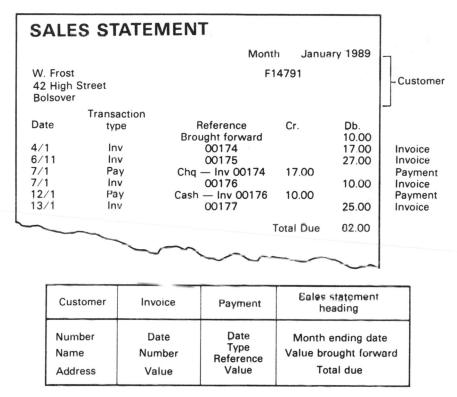

SALES STATEMENT

Month   January 1989

W. Frost   F14791   Customer
42 High Street
Bolsover

Date	Transaction type	Reference	Cr.	Db.	
		Brought forward		10.00	
4/1	Inv	00174		17.00	Invoice
6/11	Inv	00175		27.00	Invoice
7/1	Pay	Chq — Inv 00174	17.00		Payment
7/1	Inv	00176		10.00	Invoice
12/1	Pay	Cash — Inv 00176	10.00		Payment
13/1	Inv	00177		25.00	Invoice
		Total Due		02.00	

Customer	Invoice	Payment	Sales statement heading
Number	Date	Date	Month ending date
Name	Number	Type Reference	Value brought forward
Address	Value	Value	Total due

Fig. 4.34

The logical data store entity cross reference

The next step is to decompose the entity model into logical data stores. The topology of the diagram often helps with this process. A logical data store comprises the entities which form logical groups. A logical group often represents a series of entities referenced by a single process. A complementary method of arriving at logical groups is to examine the time relationships within the entities.

Consider the entity model, Figure 4.20, for the X-ray management system.

Patient records, their X-ray films and their reports are all stored until the patient death is recorded, whereas data on specific X-ray requests and appointments is transient data, only stored until the appointment including a new report is complete. Logically, three data stores exist:

DATA STORE/ENTITY X REF

SYSTEM: X-RAY MANAGEMENT	DATE:
AUTHOR: G. CUTTS	PAGE: 1 of 1

PHYSICAL/LOGICAL

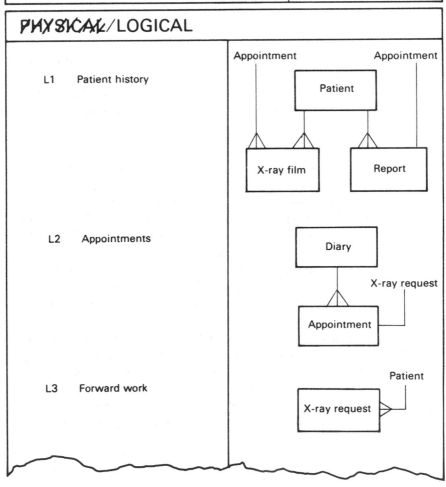

L1   Patient history

L2   Appointments

L3   Forward work

Fig. 4.35

logical data store 1 comprising patient, X-ray film and report entities; logical data store 2 comprising the diary and appointment entities; and logical data store 3, the X-ray request entities.

A cross reference can now be simply constructed between logical data stores and entities. Figure 4.35 shows such a cross reference for the X-ray management system.

The choice of the logical data store names should reflect the basic content of the data store. Logical data store 1 contains a patient history of X-ray films and reports, logical data store 2 represents all appointments, and logical data store 3 represents all the outstanding requests for X-ray, the forward load.

The decomposition of the entity model into logical data stores requires a good knowledge of the entity model. It is, however, a process which requires the application of common sense rather than absolute rigour. Any decomposition which can be argued to be a logical decomposition is correct. Figure 4.36 shows an entity model with two possible decompositions; both are correct.

The first decomposition, shown in figure 4.37, places the emphasis on accounting, with order records being close to customer records. Invoice records and stock replenishment order records are similarly close to supplier records.

The second decomposition, shown in figure 4.38, places the emphasis on stock with current stock, orders, and future stock, represented by stock replenishment orders being close. Note that in both cases the links via relationships are preserved.

The justification for the creation of logical data stores stems from the requirement to have readable data flow diagrams. Since very many systems are implemented using relational databases it might be argued that there should be one data store entitled the database, or one data store for each entity. In the first case there is a danger of hiding information and in the second case there is a danger of overloading the data flow diagrams.

Good communication with the user is the objective of data flow diagrams; common sense must therefore influence the logical division of the entity model into data stores.

Task 1.4 is now complete and two cross references have been produced. The physical data store entity cross reference provided a degree of high level proving of the DFDs and the entity model and the means by which entity names could be entered on to the current physical

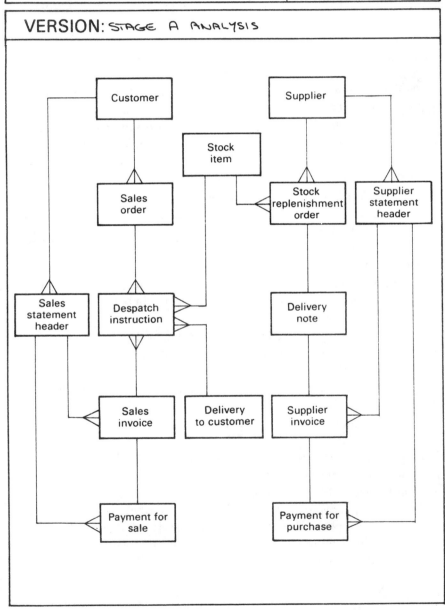

Fig. 4.36

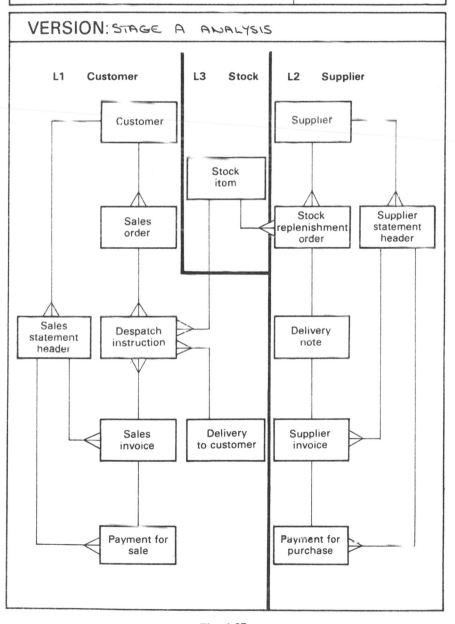

ENTITY MODEL

SYSTEM: ABC	DATE:
AUTHOR: G. CUTTS	PAGE: 1 of 1

VERSION: STAGE A ANALYSIS

L1 Customer   **L3 Stock**   **L2 Supplier**

Customer

Supplier

Stock item

Sales order

Stock replenishment order

Supplier statement header

Sales statement header

Despatch instruction

Delivery note

Sales invoice

Delivery to customer

Supplier invoice

Payment for sale

Payment for purchase

Fig. 4.37

ENTITY MODEL

SYSTEM: ABC	DATE:
AUTHOR: G. CUTTS	PAGE: 1 of 1

VERSION: STAGE A ANALYSIS

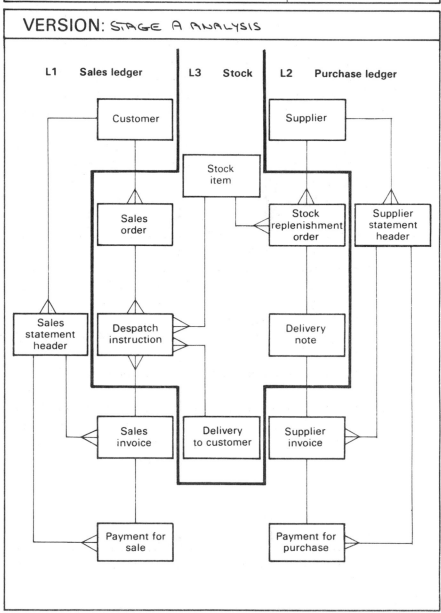

L1   Sales ledger   **L3   Stock**   **L2   Purchase ledger**

Fig. 4.38

DFDs. The logical data store entity cross reference provides one of the inputs to task 1.5.

4.3.2 Task 1.5: Create the current logical data flow diagrams

Data flow diagrams were used in task 1.2 to provide the user's view of the current system. The diagrams showed how the current system works. They were drawn to assist the analyst to understand the functions, data stores and data flows within the current system.

The construction of current physical DFDs also provided an insight into the problems within the system since the current physical diagrams should model the anomalies of the current system. Anomalies such as redundant data stores and redundant processing should have been easily identified.

The goal of system analysis is to provide a *logical* requirements specification derived from solutions to the problems, the new requirements and the *logical* model of the current system. Therefore the construction of *logical* data flow diagrams to complement the entity model is the necessary next step.

The current logical DFDs will provide a model of what the current system accomplishes, not how it is accomplished. It provides a model of the underlying functions and data, independent of the current implementation. The task of logicalising the DFDs often leads to further identification and clarification of potential problem areas.

Four steps are required to convert the current physical DFDs into the current logical DFDs. The lowest level diagrams should be converted first.

Step 1 removes physical implementation constraints from the data flows, along with their physical state; step 2 removes time dependencies associated with the current implementation; step 3 removes functions which only serve the current implementation and examines the remaining functions to remove the physical associations; step 4 incorporates the logical data stores on to the DFDs already modified by steps 1 to 3.

Logicalisation of the data flows

The first and perhaps easiest procedure is to remove all physical references on the data flows. For example, if a request document is known as the pink request form, the words 'pink' and 'form' may be deleted. The pink form represents the current implementation of a request; logically what is required is a request. The data flow label 'pink request form' is changed to 'request'. This also removes the physical format, that is a form.

This change will ensure that when the new system is implemented all possible new physical forms for a request will be considered. A request could be made by using a request form, via a screen and keyboard, or perhaps by a voice input terminal. The design of the new system should not be influenced by the current physical implementation, that is 'the pink form'. Similarly, the state of a document or set of documents should be removed. States such as 'entered on to magnetic tape' represent current physical manifestations of the data flow.

Listed below are some transformations of data flows from their physical descriptions to logical descriptions. The descriptions are transformed to reflect what the data content is, not how it is presented.

Physical	*Logical*
Pink top copy	Order
Works docket	Order
Re-order card	Request for stock replenishment
Price ticket	Price
X-ray request	X-ray request
Annotated delivery note	Quantity of parts under delivered

Step 1, therefore, is to examine each data flow, suitably renaming those which indicate a physical form or state within the name. New names should be chosen to accurately describe the data on the documents. Data flows to and from data stores will already be named with their logical entity names and need no action at this time.

Note that now the physical format of the 'pink top copy' and the 'works docket' have been removed, the data contained on the documents is revealed to be identical: 'the order'. If both of these documents are input to a single function, one of them must be redundant. The final activity within step 1 is to remove redundant data flows.

Removal of physical time dependencies

Delay is often built into physical implementations to enable efficient processing. If orders are received by telephone at a steady rate

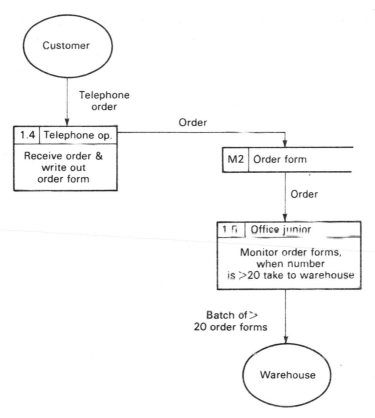

Fig. 4.39

throughout the day, it might be inefficient to take each order to the warehouse as it is received. A solution might be to record orders until twenty have been received, then to take them to the warehouse. A temporary file of orders is therefore constructed by the current physical implementation. Figure 4.39 shows the appropriate section of the current physical DFD.

Data store M2, order form, is only required to save the office junior from constantly having to take forms to the warehouse. Logically what is required is the receipt of an order via the telephone and the data being made available to warehouse staff. Data store M2 builds in a time dependency for purely physical considerations. It should therefore be deleted from the current logical DFD.

Care should be taken to ensure that data stores which do introduce

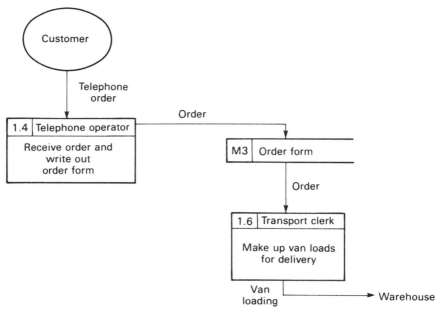

Fig. 4.40

time delays for valid reasons are not deleted. Figure 4.40 shows such a situation. In this case, the function 'make up van loads' cannot be implemented without all orders for a given route or day's delivery, depending upon the van loading algorithm. This data store introduces a time dependency but it should not be deleted as it is logically necessary.

Logicalisation of the functions

The process of removing physical references from the data flows provides a first method of identification of purely physical functions. Many physical functions serve only to transform documents with identical data content. These functions will become obvious (see Figure 4.41).

With the 'pink top copy' and the 'works docket' renamed with their data content 'order', function 1.7, 'create works order documentation', is logically redundant and should therefore be deleted.

Figure 4.39 showed a function 'monitor order forms, when number >20, take to warehouse'. This function does not carry out any logical

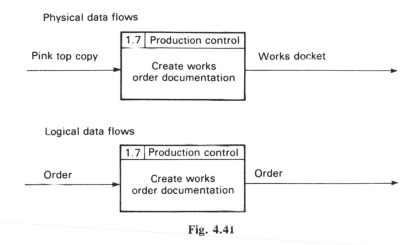

Fig. 4.41

processing; it is purely physical. Again this function should be deleted and not carried forward to the logical DFD.

The change of state of a data flow by a function often reveals further purely physical processes. The function 'enter orders to magnetic tape' is an example of a physical function which performs nothing more than a state change, which is not logical processing; it should not be carried forward to the logical DFDs.

Finally, all of the remaining functions should be renamed, if necessary, to accurately describe the logical processing. Some examples are given below:

Function descriptions

Physical	*Logical*
Write out order on pink form	Receive order
Update ledger card to reflect the payment of an invoice	Allocate payment to invoice
Type order report	Produce order analysis

Create the current logical data flow diagrams

The current physical DFDs have so far been modified by renaming the data flows, except those to and from data stores, by the removal of time dependencies and by the removal or the renaming of functions. The current logical DFDs can now be created by removing the physical data stores and incorporating the logical data stores.

This is a relatively simple activity. First, the data flows to and from data stores should be examined. These data flows were named using one or more entity names. By reference to the logical data store entity cross reference produced by task 1.4, the logical data store containing the named entity can be established. The appropriate logical data store can therefore be entered on to the logical DFD to provide a source or destination for the data flow.

Secondly, all the data flows not contributing inputs or outputs to the system should be examined. Frequently the logical name chosen for a data flow is an entity name. This data flow should be redrawn to reflect an insert into the appropriate logical data store and a subsequent read of the data store, replacing the direct flow of data between the functions. Figure 4.42 shows examples of both types of modification.

Physical data stores M2 and M3 have been replaced by one logical data store L1, order. The data store then serves both functions 1.6 and 1.7. Function 1.5 has been modified.

Note also that the responsibility section of the function rectangle is left blank on the logical DFDs.

The final step within task 1.5 is the revision and review of the newly-produced current logical DFDs.

The logicalisation task should be undertaken on the lowest level of DFD produced. The changes introduced on the lower levels should be reflected level by level on the higher levels. This will generally require the higher levels to be completely redrawn using the boundaries of the lower levels as function rectangles. Higher levels are constructed bottom up.

The removal of functions as a result of logicalisation often reduces a DFD to very few functions or even a single function. This means that the decomposition may require a second examination. Lower level DFDs may be combined and the appropriate changes made at higher levels. In this way a new set of DFDs is produced incorporating all levels of decomposition.

The current logical DFDs provide a user's view of what the system currently does, not how the system currently operates. It shows logical data flows, logical data stores and logical functions. This set of DFDs will form the basis of the requirements specification.

Physical DFD

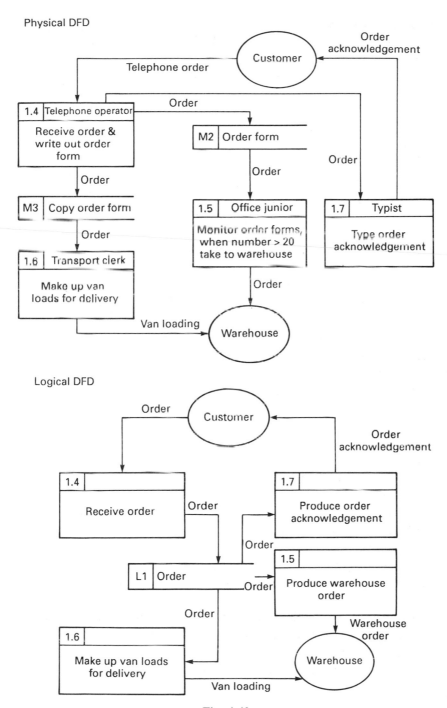

Logical DFD

Fig. 4.42

PROBLEM/REQUIREMENTS LIST

SYSTEM: *X-RAY MANAGEMENT*	DATE:
AUTHOR: *G. CUTTS*	PAGE: *1* of *1*

No.	Problem/requirement	Init.	Solution reference
1.	X-ray films and reports can be located in several places. This makes tracing of films and reports difficult in emergency situations. (Problem)	File Clerk	
2.	There is no system to ensure patients are seen by radiographers in appointment sequence. The sequence tends to be arrival sequence. (Problem)	Radio-grapher	
3.	A report showing the number and type of X-ray films produced each day. (Requirement)	Radiologist	
4.	A facility to reproduce efficiently standard reports such as 'no fracture seen'. (Requirement)	Radiologist	

Fig. 4.43

Summary of task 1.5

The current logical DFDs are a vital part of the documentation and the methodology. These DFDs will be used to develop, in part 2, the required logical DFDs. This set of diagrams represents a major part of the requirements specification, one of the vital documents in any development project.

Task 1.5 comprises four simple steps:

Step 1: logicalise the data flows
Step 2: remove physical time dependencies
Step 3: logicalise the functions
Step 4: logicalise the data stores

4.4 SECTION 3: CREATION OF THE PROBLEMS AND REQUIREMENTS LIST

4.4.1 Task 1.6: Create the problems and requirements list

This final task in part 1 produces the third output from the part: the problems and requirements list.

The list is compiled by reference to the investigation notes, and by analysis of the DFDs and entity models. A simple format is used for the list (see Figure 4.43). The format includes a description of the problem or requirement, the initials of the originator of the problem or requirement and a solution reference. At this stage the solution reference is not used; it will be added when the required system has been specified, to show that solutions have been incorporated into the new system.

Problems with the current system and requirements for the new system should be part of any information gathering techniques used during task 1.1. Interviewing is perhaps the most productive technique for establishing problems and requirements. Problems are also revealed during the construction of the DFDs and the entity model, particularly during the several tasks forming the sequence of events leading to the final current logical DFDs.

An example of a problem revealed during modelling could be the number of separate instances of similar or identical data stores. Some of these problems can be removed during the construction of the current logical DFDs; these are associated with physical problems. Logical problems will not be removed; these need to be recorded on the problems and requirements list so that in part 2 they can be removed when the logic is optimised.

Figure 4.43 shows some examples of problems and requirements for the X-ray system.

4.5 PART 1 SUMMARY

Part 1 comprises perhaps the most difficult set of tasks. They are all concerned with gaining a thorough understanding of the current system. This understanding is researched and documented to form the basis of the vital document, the specification of requirements to be produced during part 2. One of the common problems of systems analysis is the temptation to become involved in minute detail before a good understanding of the system is obtained. The methodology described for analysis in this chapter attempts to prevent this descent into detail by providing top down modelling techniques.

It is vital to part 1 that it is the analysis stage: the stage where the understanding is reached by an iterative process of investigate and model. The initial models will, therefore, not be correct; they will represent the best understanding at the point of modelling. Task 1.1, investigation of the current system, and tasks 1.2 and 1.3, the creation of the current physical DFDs and the current entity model, should be regarded as an iterative set. High level investigation should be followed by a first attempt at modelling, then detail should be added to the models by subsequent iterations of investigate and model. The modelling processes should progress until both the current physical DFDs and the current entity model have been agreed with the user.

Task 1.4, creation of the data store entity cross references, contributes towards task 1.5 where the current physical DFDs are converted, by a series of simple steps, to current logical DFDs. The diagrams are logicalised to reflect what the system does, not how the system is currently implemented. The activity of extracting the logic of the existing system is one of the major benefits of the methodology.

The final task in part 1, task 1.6, creates a list of problems with the existing system and a list of requirements for the new system. The problems and requirements are extracted from the investigation notes and from the DFDs and entity model. The analysis of the investigation notes to produce the models will reveal problems with the current system such as duplicate data stores and processing.

Part 1 is complete when the user review of the output is completed and agreed. The output documents are a set of current logical DFDs, an entity model and a list of problems and requirements.

Chapter 5
Specification of Requirements –
Outline Business Specification

5.1 INTRODUCTION

Chapters 5 and 6 describe the second part of structured systems analysis and design, the specification of requirements, shown in Figure 5.1. This part uses the logical models produced by part 1, together with the problems and requirements list, to produce a business requirements specification. This will specify what the new system is required to do and how it should be implemented.

Tasks 2.1 and 2.2, creation of the required logical DFDs and the required entity model, provide the highest level specification. They may be created by modification of the current logical DFDs and the current entity model. Modifications are made to reflect the requirements listed during part 1, task 1.6, and to provide solutions to the problems listed as part of the same task. The required logical DFDs and the required entity model may alternatively be created from the list of requirements and from the detailed knowledge of what the current system accomplishes without reference to the current logical DFDs. Tasks 2.1 and 2.2 are often linked under the heading of optimisation, improvement and development of the logical model.

An outline business specification is produced before the remaining tasks add detail to the model, as well as providing a third view of it in the form of an entity function matrix and entity life histories. These tasks document the function descriptions, the entity descriptions, the input and output and data flow descriptions, the data dictionary and the on-line dialogue as well as providing the third view of the system.

The full specification of requirements comprises:
(1) An outline business specification comprising
  - the required physical data flow diagrams
  - the required entity model
  - the design constraints

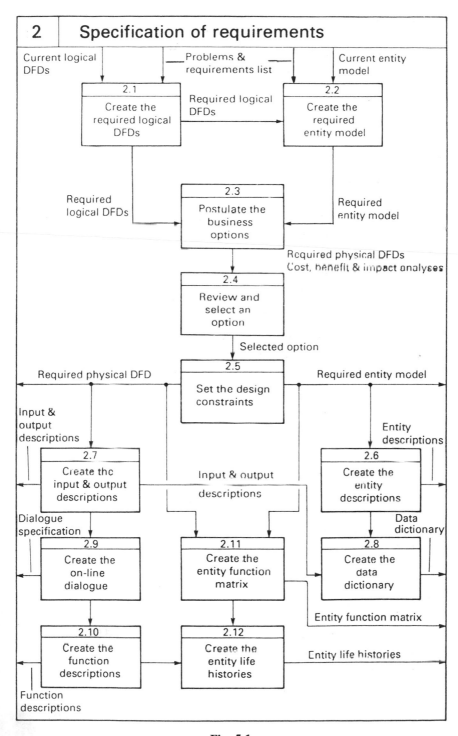

Fig. 5.1

(2) A detailed business specification comprising
  - the entity descriptions
  - the input and output descriptions
  - the data dictionary
  - the on-line dialogue specification
  - the function descriptions
(3) A validation of the specification comprising
  - the entity function matrix
  - the entity life histories

Part 3, logical data design, will use the entity model, entity descriptions and input and output descriptions to produce the final logical data design. Part 4, logical process design, will use the required physical DFDs, the entity function matrix, the entity life histories, the entity descriptions, input and output descriptions, the function descriptions, the on-line dialogue specification and the logical data design to produce the logical process catalogue and a set of logical process outlines.

Part 2 commences by optimisation, improvement and development of the outline business specification and concludes by providing a detailed requirements specification. Chapter 5 describes the creation of the outline business specification. Chapter 6 describes the creation of the detailed business specification and the validation of the specification.

5.2 CREATION OF THE REQUIRED LOGICAL MODEL

5.2.1 Task 2.1: Create the required logical data flow diagrams

Structured systems analysis and design made considerable use of DFDs in part 1; they were used to provide a view of the current system that was understandable by the user. They provided a method of testing the completeness of understanding against the observed actual system. Logical DFDs provided a method of documenting what the system accomplished, an abstraction of the logic of the current system.

Task 2.1 will again use DFDs to document logic – the logic of the required system: what the new system is required to accomplish. The completeness of understanding against the requirements for the new system is tested, instead of testing the completeness of understanding against the actual system.

DATA FLOW DIAGRAM

SYSTEM: CHAPTER 5 EXAMPLE	DATE:
AUTHOR: G. CUTTS	PAGE: 1 of 1

LEVEL: 1	~~CURRENT~~/REQ.	~~PHYS.~~/LOGICAL

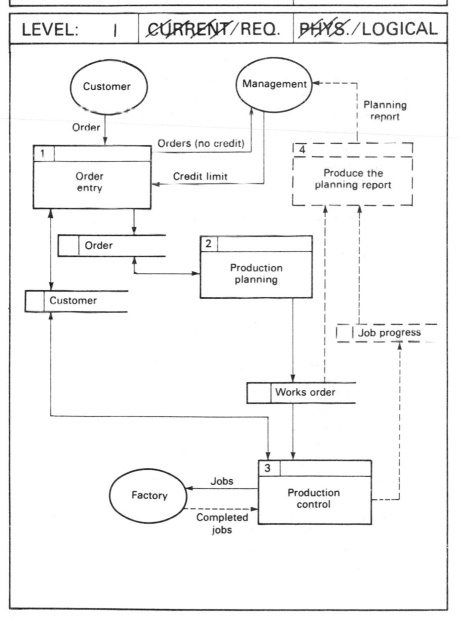

Fig. 5.2

Data flow diagrams still provide a user's view of the system. In this case, the required logical DFDs provide a view of what the new system will accomplish, not, at this stage, how it will be implemented.

Task 2.1 creates the required logical DFDs. These may be produced by modification of the current DFDs. Alternatively, they may be produced by consideration of the existing and new requirements without reference to the logic of the current system. The objective of task 2.1 is to eliminate the problems associated with the current system and to incorporate the requirements for the new system.

Figure 5.2 shows a current logical DFD with modifications to provide a new management planning report. It is, therefore, a required logical DFD. The modifications are shown using dotted structures.

The modifications comprise a new function, new data flows, a new data store and revision of an existing function. Modifications will typically comprise new functions and data stores, and revision of existing data stores and functions. Each of these modifications may result in new or revised data flows.

Figures 5.3 and 5.5 show a complete DFD before and after modification. Figure 5.3 shows the current view and Figure 5.5 the required view, after modification. Modifications were made to provide solutions to the problems and to incorporate the requirements shown in Figure 5.4.

Note that the problems and requirements list has been updated to include solution references.

Problem 1 is solved by the incorporation of a new data store, L4, by revision to functions 2 and 3 and by a new function 5. Function 3, stock check, is enhanced to include the insertion of the delivery note into a new data store L4, delivery. Function 5, record delivery success, simply modifies the delivery inserted during function 3 to indicate that a satisfactory delivery has taken place. Finally, function 2 requires enhancement to read the new data store, to access delivery information and to include this information on the management report.

Requirements 2 and 3 both require function 4, produce invoice, to be amended.

The invoice price depends upon both the customer and the product. A new entity called price is therefore required as a link between customers and products. The entity price should be added to the entity model. A new logical data store should be created for the entity, or the new entity should be added into an existing logical data store. In Figure

DATA FLOW DIAGRAM

SYSTEM: EXAMPLE	DATE:
AUTHOR: G. CUTTS	PAGE: 1 of 1

LEVEL: 1	CURRENT/~~REQ~~ ~~PHYS~~//LOGICAL

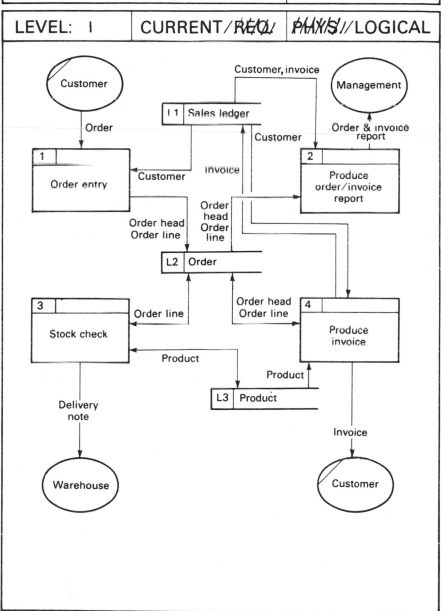

Fig. 5.3

PROBLEM/REQUIREMENTS LIST

SYSTEM: EXAMPLE	DATE:
AUTHOR: G. CUTTS	PAGE: 1 of 1

No.	Problem/requirement	Init.	Solution reference
1.	Management require the order/invoice report extended to include details of the delivery note produced for each order line. The report should also include evidence that the delivery has taken place. This is to overcome the problem of lost deliveries. There is evidence of goods being lost in transit. Two copies of the delivery note should be produced, one copy being signed by the customer and returned to the company by the delivery driver.	Sales Manager	Functions 2 3 & 5 Data store L4
2.	It is required to introduce a flexible pricing structure whereby the price of each product may vary according to a customer.	Sales Manager	Additional Entity Price
3.	Invoices should not be generated until a satisfactory delivery has taken place. The current system allows invoices to be generated after production of a delivery note which does not guarantee despatch from the warehouse.	Chief Accountant	Function 4

Fig. 5.4

DATA FLOW DIAGRAM

SYSTEM: EXAMPLE	DATE:
AUTHOR: G. CUTTS	PAGE: 1 of 1

LEVEL: 1	~~CURRENT~~/REQ.	~~PHYS~~./LOGICAL

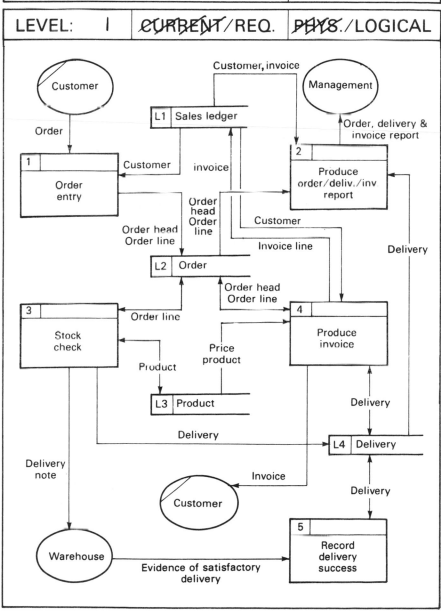

Fig. 5.5

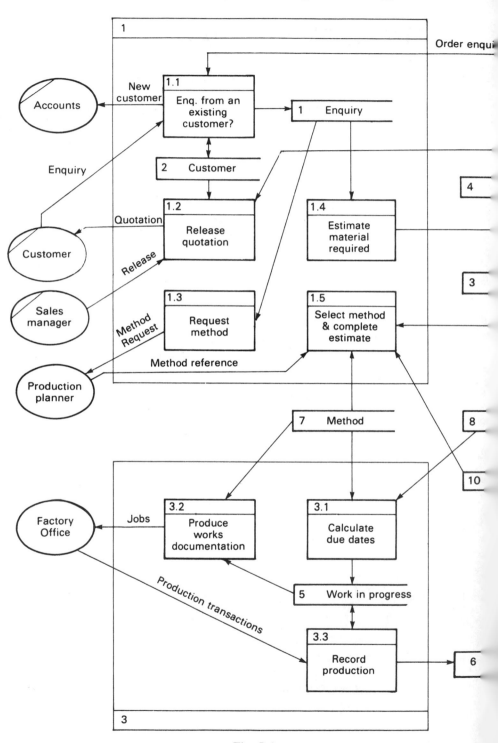

Fig. 5.6

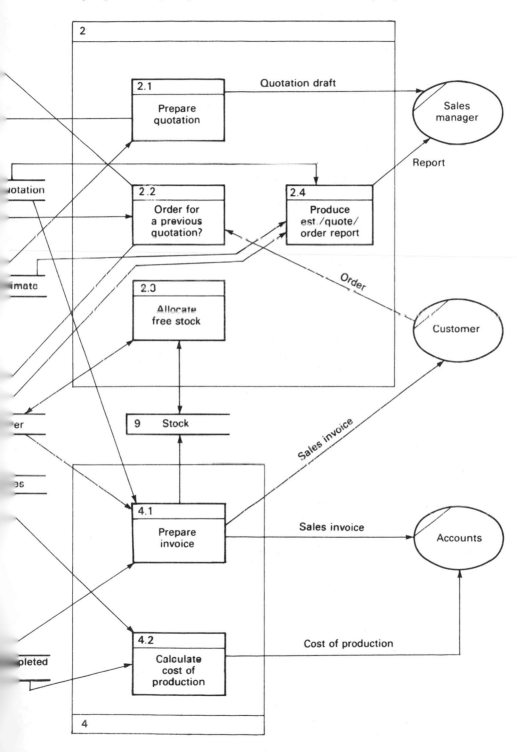

5.5, the entity price has been incorporated into logical data store L3, product. The invoice price is obtained from the price entity within the logical data store L3 (see Figure 5.5).

Function 4, produce invoice, references data store L4, delivery. Deliveries are marked to indicate a satisfactory delivery has taken place by the new function 5. Invoices may only be produced for deliveries so marked, thus satisfying requirement 3 on the problems and requirements list.

Figure 5.5 shows a required logical DFD which has been developed by modification of the current logical DFD. The decomposition into functions which was considered when the current DFDs were developed still exists. Some aspects from the current physical system, for example current departments or sections as functions, still exist on the required logical DFDs.

This may be a major disadvantage with the method of developing required logical DFDs. It may also be a major advantage since the final design may preserve an existing department's or section's major functional responsibilities. A statement to that effect in the terms of reference is not uncommon.

Using the current logical model as a base for the required logical model may also be a major disadvantage. Responsibilities may have been allocated in the current system to serve operational needs. The decomposition into departments and sections may exhibit little logic.

The alternative method of developing the required logical DFDs allows the analyst to commence with a list of requirements and to develop a set of DFDs to satisfy them. The DFDs need not be influenced by any aspects of the current system.

The decomposition of DFDs into functions and sub-functions can be based completely according to the rules of coupling and cohesion. Each decomposition must seek to produce highly cohesive functions which represent one set of system related activities, loosely coupled to other functions, that is with a minimum number of inter-function data flows.

One method of obtaining high cohesion and loose coupling is by bottom up development. A DFD should be produced at the lowest level showing all the primitive functions. Functions should then be grouped to form the levels of decomposition required to assist the understandability of DFDs. Figure 5.6 shows a large DFD divided into functional areas.

The required logical DFDs may, therefore, be developed without being influenced by the current system. If the system being developed is entirely new and no current system exists, then this may be the only method of developing the required logical DFDs. They are transformed into required physical DFDs. Several required physical DFD sets may be produced to represent different implementation or business options: tasks 2.3 to 2.5.

For many systems, the required logical DFDs can be eliminated. The required physical DFDs can be created to satisfy the requirements. There is some merging of the tasks.

5.2.2 Task 2.2: Create the required entity model

The entity model which resulted from part 1 modelled the generic underlying structure of the data and data relationships associated with the current system. The creation of the required logical DFDs during the previous task may well have introduced new entities into the model. It is also possible for entity descriptions to be modified and for entities to be deleted. The addition, deletion and modification of entities needs to be reflected in the entity model. However, the changes made to the DFDs are not entity additions, deletions and modifications; they are additions, deletions and modifications to logical data stores on the DFDs.

Part 1, task 1.4 transformed the entity model into a set of logical data stores. What is required now is the reverse process: the transformation of the new logical data stores into a revision of the entity model. The data store entity cross reference will require revision to include the entities created. The new entities should then be added to the entity model.

Appropriate relationships should be generated and a revised entity model created for further refinement in stage B. The new data store added to Figure 5.3, delivery, shown in figure 5.5, results in modification to the data store entity cross reference and the entity model. Figure 5.7 shows the original entity model and Figure 5.8 the revised entity model.

There now exists a price entity between customer and product. A customer has many prices, one for each product, and a product has many prices, one for each customer. A new entity, price, accessed by function 4, produce invoice, is introduced.

A new entity, delivery, has also been introduced. Customers receive many deliveries. Each delivery comprises many order lines; however, an order line can only have one delivery (no part deliveries) and be to one customer.

ENTITY MODEL

SYSTEM: CHAPTER 5 EXAMPLE	DATE:
AUTHOR: G. CUTTS	PAGE: 1 of 1

VERSION: STAGE 2

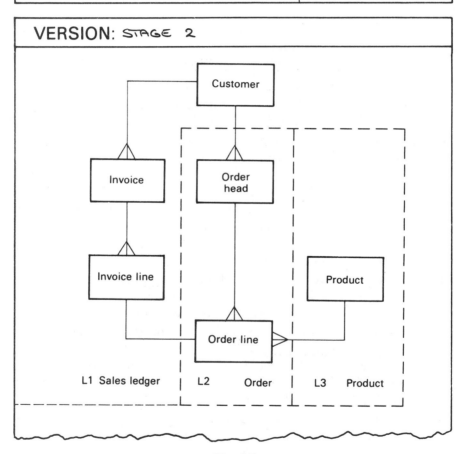

Fig. 5.7

ENTITY MODEL

SYSTEM: CHAPTER 5 EXAMPLE	DATE:
AUTHOR: G. CUTTS	PAGE: I of I

VERSION: STAGE 2

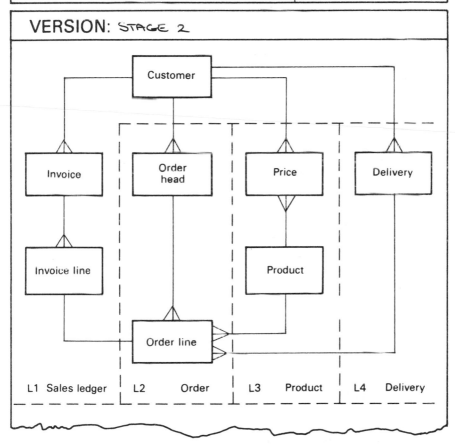

Fig. 5.8

5.3 CREATION OF THE OUTLINE BUSINESS SPECIFICATION

5.3.1 Introduction

The prime inputs are the required logical DFDs and the required entity model. However, many of the other pieces of documentation, such as the feasibility study report, will act as useful reference documents. The objective is to produce an outline business specification: a specification of how the new system will be implemented in the business.

The required logical DFDs represent a solution to the system problems as well as incorporating the logical requirements for the new system. The objective is to produce a set of physical DFDs which represent the required system, having taken due account of all the constraints including hardware, software and personnel available. A new physical system is required which offers a best fit with the requirements, the hardware, the software and the user, and is achievable within the resources available for further development and implementation.

This is the business specification. It provides a formal link between the feasibility study findings, the terms of reference for this project, the developing specification of requirements and the system to be designed and implemented.

There are three tasks within this section. Task 2.3 postulates a number of business options together with a cost benefit analysis for each option, task 2.4 makes a selection of one of the options, and task 2.5 sets some additional constraints on the design for the selected option. This option is then developed into a detailed specification.

5.3.2 Task 2.3: Postulate the business options

Figure 5.9 shows a level 1 required logical DFD. The responsibility box associated with each function was not used on the logical DFDs. This box is now used to indicate how the processing represented by each function box will be implemented. Options for the completion of the box are, therefore, words such as: computer, sales office, depot, warehouse etc. Indeed, with each of the options further detail could be added such as on-line computer or batch computer.

The initial exercise is carried out using the level 1 DFD, the highest level. It is possible that, at this level, a function might be part clerical

DATA FLOW DIAGRAM

SYSTEM: EXAMPLE	DATE:
AUTHOR: G. CUTTS	PAGE: 1 of 1

LEVEL: 1	~~CURRENT~~/REQ.	~~PHYS~~./LOGICAL

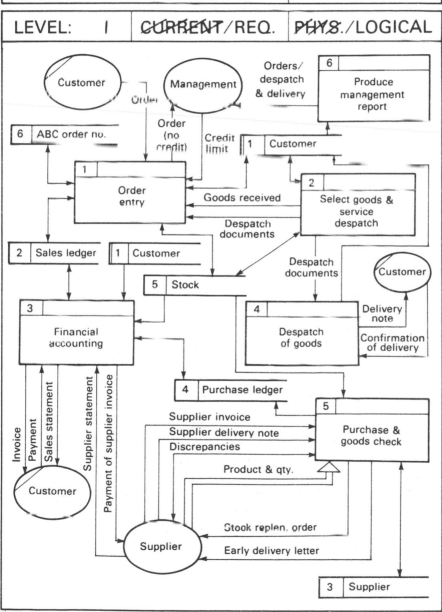

Fig. 5.9

and part computer. Entries such as warehouse/computer are acceptable with the level 2 DFD being used to indicate exactly which sub-functions are to be implemented via a computer and which sub-functions will be carried out by staff in the warehouse.

Figures 5.10, 5.11 and 5.12 show three possible business options. The option in Figure 5.10 makes the maximum use of the computer. The option in Figure 5.11 uses the computer for functions dealing with orders and customers and the option in Figure 5.12 uses the computer for financial functions only.

The computer system boundary

The boundary of the computer system can be established for each option. All functions for which the responsibility is computer, are inside the boundary. If a level 1 function has a joint computer/clerical responsibility, e.g. Figure 5.12, function 5, then the boundary may be determined by including some of the functions from the appropriate level 2 DFD.

Figure 5.13 shows option 3, Figure 5.12, in more detail. Function 5, where the responsibility was shared between computer and accounts, is decomposed to show exact responsibilities using the level 2 DFD.

All functions, and their sub-functions, marked 'computer' are considered to be inside the boundary of the required new system. The inclusion of functions within the boundary is, therefore, straightforward but the inclusion of data stores is not so easy. If a data store is excluded, then any interaction with that, now manually maintained, data store and the computer must be via input or output documents. Alternatively, if a data store is included, then any interaction with non-computer functions must be via input or output documents.

The number of data flows each data store possesses with functions inside and outside the boundary is one guide as to whether the data store should be included or excluded. A second guide is the maintenance of the data store. If the transactions which maintain a data store are within the system, then the data store should be within the system. A third guide is the problems and requirements list. If it is necessary to provide a computer data store to solve a stated problem or meet a new requirement, then the data store should be included inside the boundary.

The boundary of the computer system can now be fully agreed and

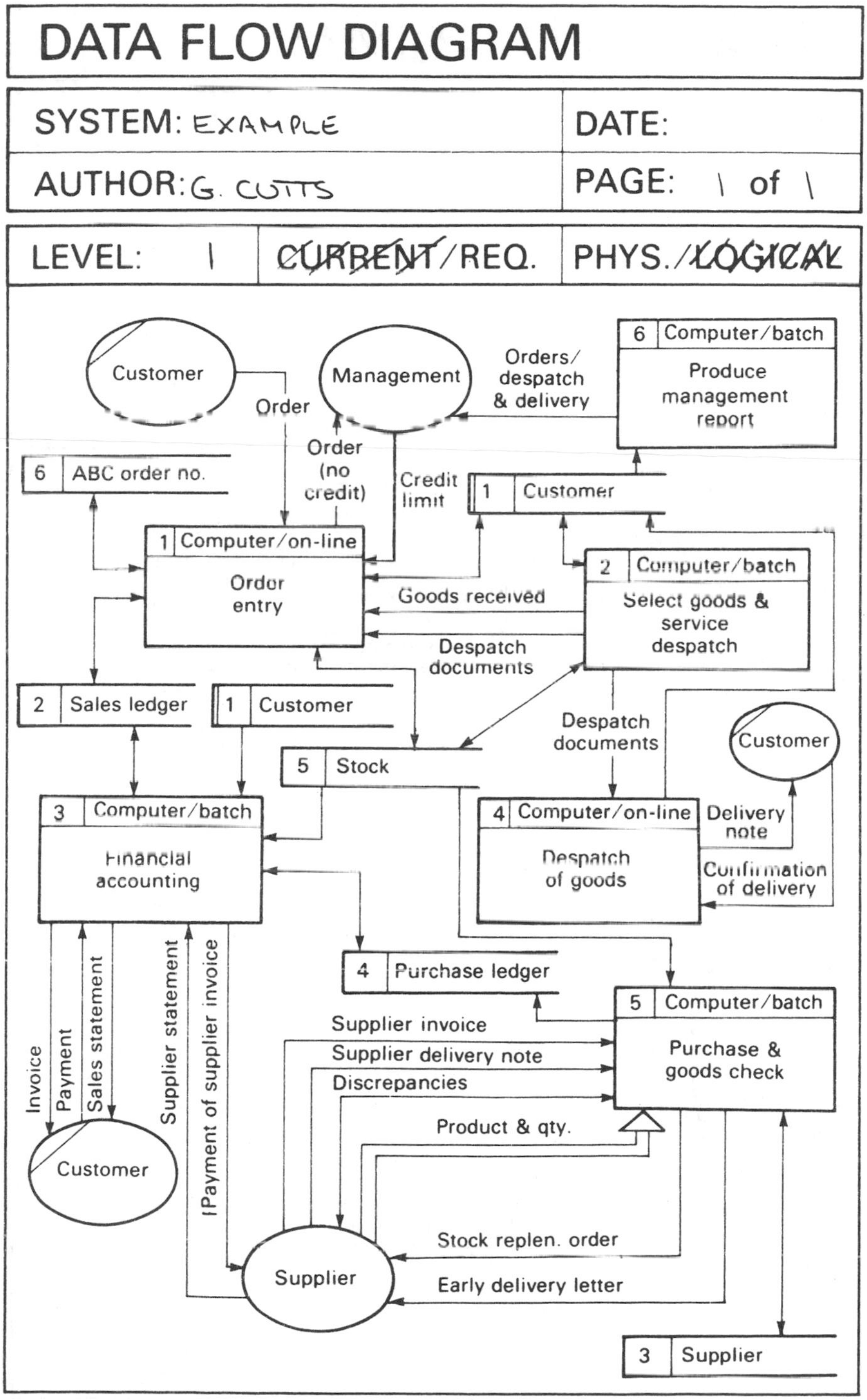

DATA FLOW DIAGRAM

SYSTEM: EXAMPLE	DATE:
AUTHOR: G. CUTTS	PAGE: 1 of 1

LEVEL: 1	~~CURRENT~~/REQ.	PHYS./~~LOGICAL~~

Fig. 5.10

DATA FLOW DIAGRAM

SYSTEM: EXAMPLE	DATE:
AUTHOR: G. CUTTS	PAGE: 1 of 1

LEVEL:	~~CURRENT~~/REQ.	PHYS./~~LOGICAL~~

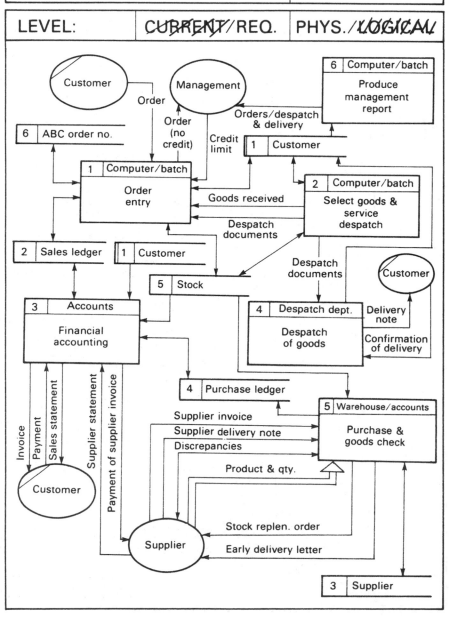

Fig. 5.11

DATA FLOW DIAGRAM

SYSTEM: EXAMPLE	DATE:
AUTHOR: G . CUTTS	PAGE:	of	

| LEVEL: | | ~~CURRENT~~/REQ. | PHYS./~~LOGICAL~~ |

Fig. 5.12

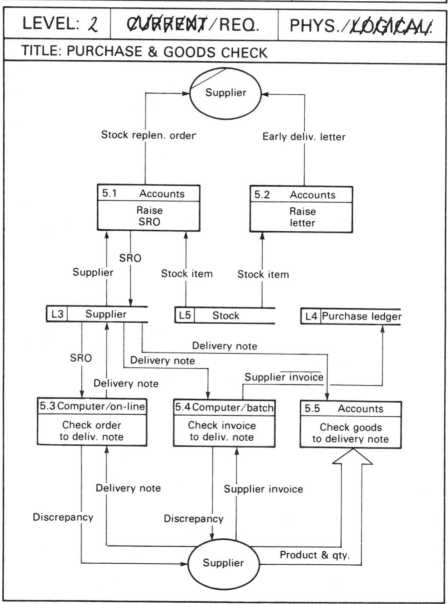

DATA FLOW DIAGRAM

| SYSTEM: *A B C* | DATE: |
| AUTHOR: *G. CUTTS* | PAGE: *1* of *1* |

| LEVEL: *2* | ~~CURRENT~~/REQ. | PHYS./~~LOGICAL~~ |

TITLE: PURCHASE & GOODS CHECK

Supplier

Stock replen. order   Early deliv. letter

| 5.1   Accounts | 5.2   Accounts |
| Raise SRO | Raise letter |

SRO
Supplier   Stock item   Stock item

| L3 | Supplier | | L5 | Stock | | L4 | Purchase ledger |

Delivery note

SRO   Delivery note

Delivery note

Supplier invoice

| 5.3 Computer/on-line | 5.4 Computer/batch | 5.5   Accounts |
| Check order to deliv. note | Check invoice to deliv. note | Check goods to delivery note |

Delivery note   Supplier invoice

Discrepancy   Discrepancy

Supplier   Product & qty.

Fig. 5.13

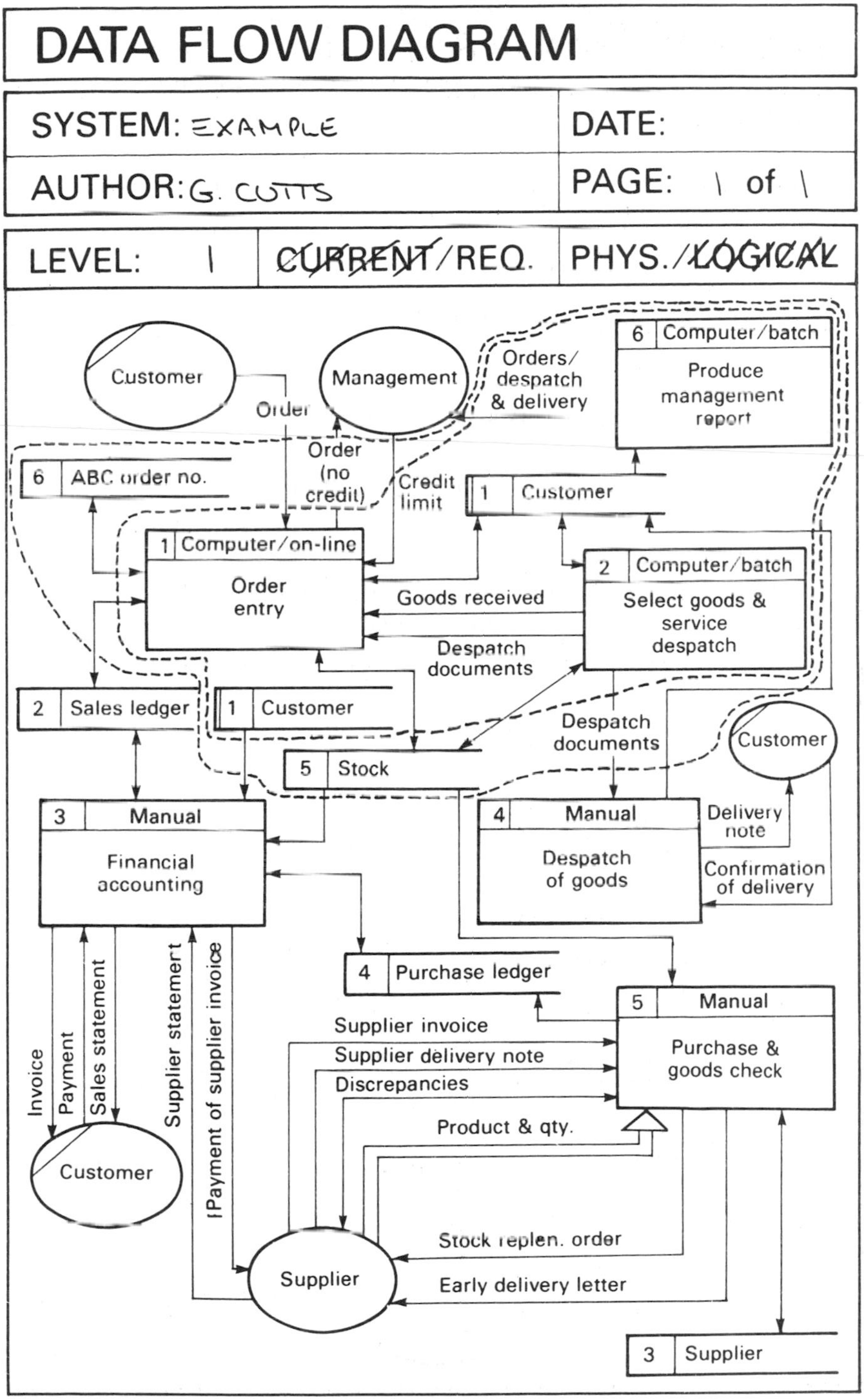

DATA FLOW DIAGRAM

SYSTEM: EXAMPLE	DATE:
AUTHOR: G. CUTTS	PAGE: 1 of 1

LEVEL: 1	~~CURRENT~~/REQ.	PHYS./~~LOGICAL~~

Fig. 5.14

the new physical functions, data stores, input and output documented. Each option will have different system functions, data stores, input and output. In general, therefore, each option will have a different impact on the problems and requirements list. The impact on the list must be documented as part of the option specification.

Figure 5.14 shows two possible boundaries. Function 5, which will be implemented as a manual process, requires information from data store 5. With the outer boundary a computer function is required to provide information, either as an on-line enquiry function or as a computer report. In this case a daily report is appropriate so a new function, produce daily low stock report, is required. This is an additional output from the system which will, of course, affect the total cost of implementing the option. This new function and output must be added to existing documentation.

Data store 6 provides the next order number to be used as a unique key for orders received. There are two options shown. Option 1, the inner boundary, would require the computer to print the last used number. At the start of the next run, the number could be input via parameter to the program. Option 2 would require the program to store the number at the completion of processing and to retrieve it at the start of the next run.

Part of the option specification is, therefore, a set of required physical DFDs. The diagrams form one of the inputs to task 2.4, option selection.

Constraints

All the required physical systems postulated must adhere to all the constraints of the project. Decisions on hardware, software, personnel, timescales, costs etc., which were not specified as part of the project terms of reference or as part of the feasibility study report, must now be taken. Some of the system constraints will have emerged during the detailed investigation and modelling tasks. Mandatory reports, mandatory on-line facilities and minimum response times are examples of constraints which emerge during detailed investigation. These constraints should be documented as part of the requirements list produced during part 1.

In general terms, however, very few projects enjoy a free choice of hardware, software, personnel, development timescales etc. These constraints represent decisions taken very often before the commencement of the project. Projects have to be implemented on currently available hardware, with existing software and personnel, and with development timescales imposed by management. This is the real environment of many projects.

Constraints which must be considered include the resources available for further development, costs and timescales, mandatory requirements, logical and physical; and the required service levels, response times and performance levels.

Implementation considerations

The responsibility box on the required physical DFDs, developed at the start of task 2.3, specified for each function how it would be implemented: on-line, batch or manual. This is just one of the considerations regarding the planned implementation. Other implementation decisions need to be taken before the detailed design stages may commence.

Decisions need to be taken regarding the hardware and software to be used. For example: 'Is a package to be used for part or all of the system?'; 'Does the software provide sophisticated database management facilities?' 'Will the system be centralised or distributed?'; 'What is the total available on-line disk capacity?'; 'What method of communication will be used?'.

Decisions taken on issues such as these will have an impact on the design stages. Decisions taken on many of these issues will also have an impact on the user. The user must, therefore, be involved in the discussions on the user interface, and on organisational and staffing issues.

The final areas for detailed consideration before any option is selected are the design and implementation methods to be employed throughout the remainder of the project. The constraints must be observed by all the postulated options. Consideration of implementation matters may well result in different decisions for each option. The effect of the decisions taken must be carefully documented and presented to task 2.4 as part of the option specification.

The option specification comprises a set of physical DFDs and a narrative addressing business issues.

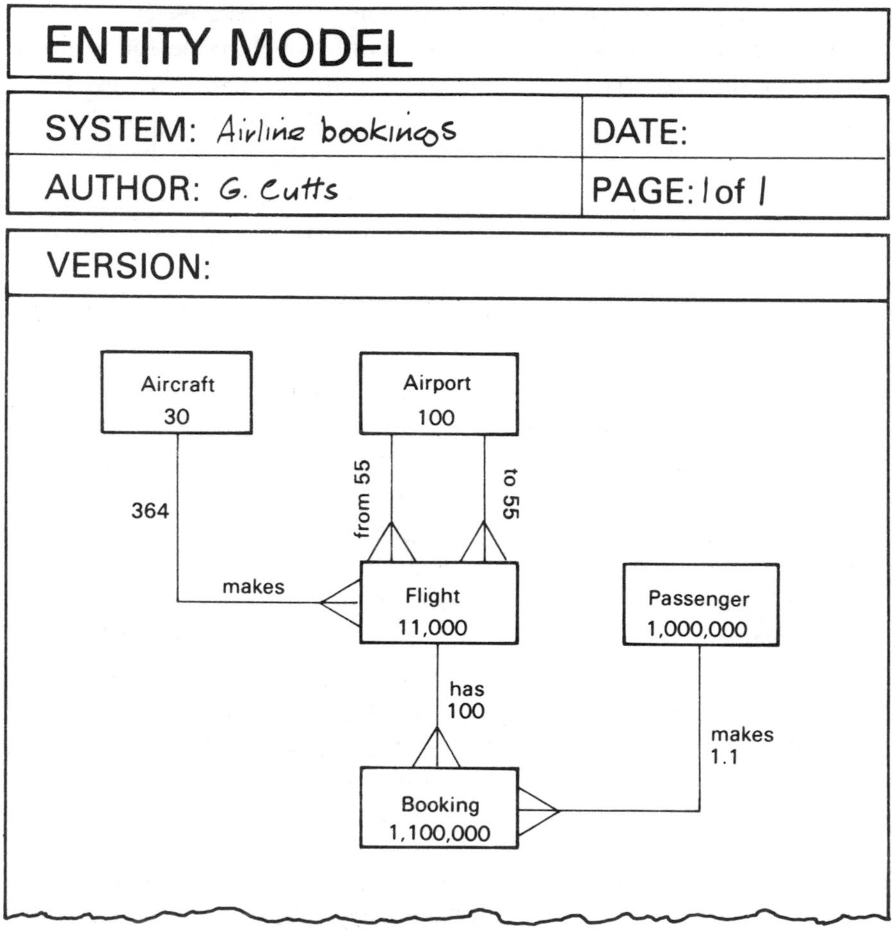

ENTITY MODEL

SYSTEM: Airline bookings	DATE:
AUTHOR: G. Cutts	PAGE: l of l

VERSION:

Fig. 5.15

Outline sizing

The selection of hardware may well be influenced by the total amount of on-line disk capacity required. An estimate of the disk capacity can be obtained by simple processing of the entity model. The entity model shown in Figure 5.15 has been annotated with figures to show the maximum number of occurrences of each entity. All data is retained in the system for six months. For example, an aircraft makes two flights per day, seven days per week. The data for twenty-six weeks is held within the system, so each aircraft has data on $2 \times 7 \times 26 = 364$ flights.

There are then approximately:

30 aircraft each making 364 flights	= 11,000 flights
100 airports each with 55 flights in and out	= 11,000 flights
11,000 flights each with 100 bookings	= 1,100,000 bookings
1,000,000 passengers each making,	
on average, 1.1 bookings	= 1,100,000 bookings

The next task is to estimate the size of each entity. For example, the passenger entity may have name and address as attributes, giving a size of 250 bytes, say.

The total disk capacity required for system data can easily be calculated:

Entity name	No. of entity occurrences	Entity size	Space required
Aircraft	30	100	3,000
Airport	100	300	30,000
Flight	11,000	100	1,100,000
Booking	1,100,000	50	55,000,000
Passenger	1,000,000	250	250,000,000
Total			306,133,000

The total space required is just over 300 Mbytes plus an allowance for system software. This may be too large. It may be necessary, therefore, to reduce the number of passenger records maintained on-line.

Supporting narrative

The options specification may require further narrative to make it acceptable to users. The physical DFDs may need to be enhanced with descriptions of the environment and the system. The environment description should address both technical and social issues. The technical descriptions should provide an overview of the hardware and software to aid the user's understanding and so support the decisions documented as part of the business considerations. The social description should provide an overview of the user's operational environment to further the user's understanding and to provide a basis for discussion of the business considerations.

Both narratives should refer to existing documentation whenever possible but they should not provide a narrative of existing documents. Those which should be referenced include the required physical DFDs, the hardware and software vendor manuals, plus any relevant internal standards such as project control standards, methods standards or performance standards.

The option specification requires two further items: an impact analysis, specifically addressing the impact on the problems and requirements list of the option, and an option cost benefit analysis.

The impact analysis should document precisely, for each problem, how the problem is overcome in the new system. Many problems are removed when the current physical DFDs are transformed to the current logical DFDs; i.e. many of the problems that users describe are associated with the current implementation. There is a tendency, however, to reintroduce problems of a physical nature when transforming the required logical DFD to the required physical DFD. Problems of duplicated data stores and physical functions emerge. It is important, therefore, to describe precisely how, in real terms, each problem is solved in the new system.

The impact analysis should also document how each of the requirements is implemented in the new system.

The ratio of computer functions to manual functions and the number of computer files will determine the costs for design and implementation of the system. They will also determine the costs of hardware, software, training, operation and future modification and maintenance. A very good estimate of the future cost of continuing the project can therefore be calculated. Further, by reference to the impact analysis and the technical and social descriptions, the benefits to the organisation of the project can be established.

A cost benefit analysis can now be undertaken for each option.

An option specification comprises a set of required physical DFDs; a narrative describing technical aspects, social aspects and implementation considerations; an impact analysis; and a cost benefit analysis.

The effort required to produce an option specification is very considerable, even if it can be based upon a good feasibility report. Two or three options only should be developed to this advanced stage. A high level selection process is required to eliminate all the options which are technically, socially or economically unacceptable. This can be done by careful study of the required physical DFDs. Six or seven options can be postulated using DFDs, with two or three being selected for detailed analysis.

5.3.3 Task 2.4: Review and select option

The selection process chooses between a number of business options, selecting one for continued development. The selection should be carried out by the user, advised and guided by the project team. Four sets of documentation assist the decision making process: the physical DFDs, the narrative, the impact analysis and the cost benefit analysis, which make up the outline business specification.

All the options should meet the constraints set out in the terms of reference and in the feasibility study report. Typical constraints would be on resources available for the development; hardware, software and staff; and on the timescale for development, say, within six months of the commencement of the project. The selection process is very much, therefore, between business options, that is:

- What will be the impact on the user's operation?
- What processing will be manual?
- What will be on line?
- What will be batch?
- Which data stores will be computer files?

One method of involving users and senior management in the selection process is to prepare a presentation. This should present the logical specification followed by overviews of each of the options. The overviews should concentrate on how, in each case, the logical specification would be implemented. This initial part of the presentation should concentrate on educating the users on the options. The second part should concentrate on providing sufficient information to allow the user, assisted by the project team, to select one of the options.

The major items to be included within the presentation are the impact analysis and the cost benefit analysis.

A possible consequence of the presentation may be that no decision is reached. The project team must then be available for follow-up advice and guidance during the decision making process. They must also be available to rework options for minor changes, to combine options and to develop further options as directed by the user senior management.

In all decision making processes, no decision is the easiest option. It is the responsibility of the project team to record the user's requests and management decisions so that the user can be guided towards the selection of an original or reworked option for further development.

5.3.4 Task 2.5: Set the design constraints

Many of the constraints on the design have been specified as part of the option specification. The new system input, output, files and functions are identified on the required physical DFDs. The hardware and software available are described in the narrative and the resources available for the design and implementation are documented as part of the cost benefit analysis. All of these constraints must be adhered to during detailed design.

However, there are other design considerations. The design must meet standards on security, recovery and privacy. It must adhere to any installation standards and it must provide for efficient audit and control. If installation standards exist, then it is sufficient to state that the design must meet those standards. If no standards exist, the required level of security, recovery and privacy must be specified before stages B and C commence.

5.3.5 Summary

This section is not a reworking of the feasibility study. It is a further investigation into the options available for implementation within the constraints already set out in the terms of reference and the feasibility study. Constraints such as the budget, the hardware and software available and the personnel available must be honoured if set out in the feasibility study documentation. All options postulated must adhere to these constraints.

It is an investigation and specification of various business options. Each option specifies a different interpretation of the logical specification of requirements.

Selection of the business option is the most important step. It involves the user in a detailed examination of the specification of requirements in the form of business options. The selection of one of the options by the user is, in effect, acceptance by the user of the specification of requirements, which contributes a major milestone in any project.

The project team can now move forward with confidence to detailed specification, design and implementation.

Chapter 6
Specification of Requirements –
Detailed Business Specification

6.1 INTRODUCTION

Chapter 6 describes sections 3 and 4 of part 2, the specification of requirements. Section 3 develops the detailed business specification and section 4 validates the specification for completeness before the design stages commence.

6.2 CREATION OF THE DETAILED BUSINESS SPECIFICATION

6.2.1 Task 2.6: Document the entity descriptions

An entity description comprises a list of attributes. For example, the customer entity may have the following attributes: name, address, delivery address, credit limit, telephone number, date of last order etc. The entity descriptions should be completed in as much detail as possible as part of the specification of requirements.

Task 2.6 should not be considered to be the sole task where entity descriptions are considered. It is the task that brings together all the data collected on attributes to produce the formal documentation. From the moment that entities are introduced in task 1.3, entity descriptions should be considered.

Examine the physical data store entity cross reference shown in Figure 6.1. Three physical data stores – credit limit, name and address, and part of sales ledger – are cross referenced with the entity, customer. The conclusion from this is that the customer entity must contain all the attributes, or selected attributes, from all of the physical data stores. The customer entity must include, as a minimum, credit limit, name and address as attributes.

This process is slightly more difficult when a physical data store is cross referenced with a number of entities. Figure 6.2 shows a sales

DATA STORE/ENTITY X REF.

SYSTEM: CHAPTER 5· EXAMPLE	DATE:
AUTHOR: G.CUTTS	PAGE: 1 of 1

PHYSICAL/~~LOGICAL~~

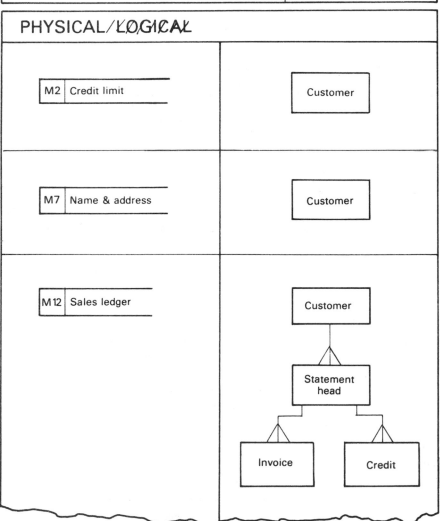

Fig. 6.1

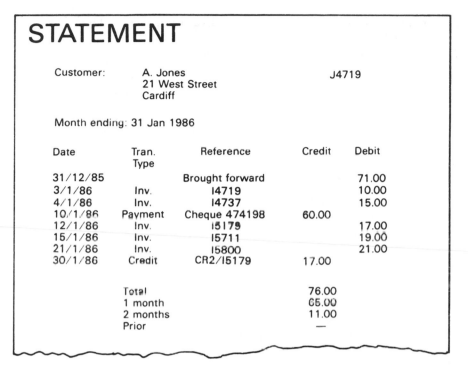

STATEMENT

Customer: A. Jones   J4719
                      21 West Street
                      Cardiff

Month ending: 31 Jan 1986

Date	Tran. Type	Reference	Credit	Debit
31/12/85		Brought forward		71.00
3/1/86	Inv.	I4719		10.00
4/1/86	Inv.	I4737		15.00
10/1/86	Payment	Cheque 474198	60.00	
12/1/86	Inv.	I5179		17.00
15/1/86	Inv.	I5711		19.00
21/1/86	Inv.	I5800		21.00
30/1/86	Credit	CR2/I5179	17.00	
	Total		76.00	
	1 month		65.00	
	2 months		11.00	
	Prior		—	

Fig. 6.2

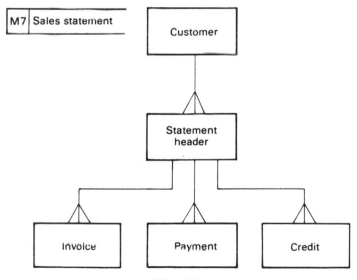

Fig. 6.3

statement. The cross reference of the data store, sales statement, with the entity model is shown in Figure 6.3.

Each attribute on the sales statement should be allocated to one of the entities. It will, therefore, become an attribute of one of the entities. The current data store items should be listed, and by reference to the data store entity cross reference, allocated to an entity. A typical allocation is shown in Figure 6.4.

This method will work for all entities identified during part 1. Entities may have been added during task 2.2 to satisfy a new requirement. In this case, an attribute list for the new entity can be created to satisfy the new requirement. Figure 5.8 included a new entity, price; its attributes might be:

- customer number
- product code
- selling price
- date of last price change.

The entity descriptions created should be in business terms. Technical detail will be added during part 5. The entity descriptions will also be validated in part 3. A typical entity description is shown in Figure 6.5.

Entity	Attribute list	
Customer	Customer	Number
		Name
		Address
Statement header	Month ending date	
	Value b/f	
	Value c/f	
	Value 1-1 month	
	Value 2-2 months	
	Value 3-prior	
Invoice	Invoice date	
	Invoice number	
	Value	
Credit	Credit date	
	Credit note number	
	Credit reference	
	Value	
Payment	Payment date	
	Payment type	
	Payment reference	
	Value	

Fig. 6.4

ENTITY DESCRIPTION

SYSTEM: *EXAMPLE*	DATE:
AUTHOR: *G.CUTTS*	PAGE: *1* of *1*

NAME: *CUSTOMER*

NARRATIVE: All customers attributes <u>not</u> related to a specific order.

Key	Data item	Format	Len.	Comment
✓	Customer Number			
	Name			
	Address			5 lines
	(Delivery Address)			5 lines
	Credit Limit			
	(Date of last order)			
	Area code			
	(Discount code)			
	(Salesman's reference)			
	Method of delivery			

	VOLUMETRICS	
	ENTITY SIZE	
	No. OF OCCURRENCES	
	TOTAL	

Fig. 6.5

172   *Structured Systems Analysis and Design Methodology*

Note that only the attribute and comment columns have entries. Round brackets (delivery address) indicate that an attribute value need not be present; the attribute is optional.

6.2.2 Task 2.7: Prototype and document the input and output descriptions

The required DFDs comprise data stores, input and output data flows, internal data flow and functions. The previous task produced the detailed documentation of the entities. The data store entity cross reference, together with the entity descriptions, effectively documents the data stores. This task documents the data flows.

For existing data flows the task is relatively simple. For example, the data flow sales statement probably corresponds to a document which is already in existence. The attributes on the sales statement can be listed, discussed with the user and documented on the appropriate form. The user may add or delete attributes from the list.

The sales statement shown in Figure 6.2 is documented and shown in Figure 6.6. Square brackets, [], show alternative attributes or groups of attributes and the vertical bar indicates a repeating group of attributes.

Many of the data flows are particular occurrences of an entity and will be named with the entity name on the DFDs. These data flows have already been documented during task 2.6. Some will comprise several entities; again, these data flows have been documented.

Task 2.7, therefore, completes the documentation of all data flows and data stores. This task is perhaps one of the most important ones. It documents, for the user and the designer, the input to the system and the output produced by the system: the user interface. It may be necessary at this stage to prototype some sample physical input forms, output reports and screen formats to assist the user specification task. Indeed, at this stage, it might be politically expedient to devote major effort to obtaining agreement on the exact requirements for input and output, via the design of the physical system's input and output.

Task 2.7 can become one of the more visible tasks to the user, where the user can become deeply involved. The user will eventually be asked to agree the requirements specification. The approach of the user will often be to satisfy themselves that the input and output is satisfactory, leaving the detail to the system's analysts and designers.

Prototyping of the system input and output is an excellent way to

INPUT OUTPUT DESCRIPTION

SYSTEM: EXAMPLE	DATE:
AUTHOR: G. CUTTS	PAGE: 1 of 1

NAME: SALES STATEMENT	

NARRATIVE: This output documents all invoices, credit notes, and payments posted to the sales ledger during the month.

Key	Data item	Format	Len.	Comment
	Customer number			
	Name			
	Address			
	Statement date			
	Value brought forward			Previous month's carried forward
	Value carried forward			To next month
	Value owing 1 month			
	Value owing 2 months			
	Value owing prior			
	Transaction type			Invoice, payment or credit
	Date of transaction			
	Invoice number			
	Payment type, reference			
	Credit note number, reference			
	Value			
	Credit/debit			

Fig. 6.6

bring the system to life for the user. The visualisation of input and output descriptions involves the user in screen and layout design, ensuring the user's understanding and acceptance of the interface. Human/machine interaction is emerging as a valid area of interest and research for computer professionals and psychologists, leading to a better understanding of factors such as how much information should be displayed on a screen and the number of items of data a user can memorise from screen to screen. Many new texts on prototyping and human/machine interaction have been published recently.

Input and output prototyping is linked to on-line dialogue design, task 2.9. These two tasks attempt to demonstrate to the user how the system will operate. The user should be deeply involved with these final tasks within the detailed specification.

Prototyping has become more common with the introduction of software tools such as screen painters, report generators, fourth generation systems and advanced operating systems. Software tools are required to enable rapid production and amendment of screen formats and report layouts, ensuring incremental development to an agreed position within acceptable timescales.

6.2.3 Task 2.8: Create the data dictionary

The documentation of the entities and the data flows provides a list of attributes or data items to be included in the data dictionary. At this stage it may only be possible to enter the attribute's name into the data dictionary, designating the entry as 'undefined'. The final definition of attributes results from further discussion with the user, particularly during prototyping and physical design.

The attribute's data dictionary entry should comprise the following, as a minimum entry.

Data/User name	the name used on user documentation such as the DFDs and entity models
Short name	the name to be used within the computer system
Type	character, numeric, etc.
Length	in bytes, say
Format	number of decimal places, mix of alpha and numeric, etc.
Characteristics	signed, rounded, packed, justified, etc.
Range	e.g. 0 to 99

Validation   check digit etc.

Plus a description of the role and purpose of the attribute.

The data dictionary is one of the important items of documentation made available to the implementation team.

6.2.4 Task 2.9: Prototype and create and on-line dialogue specification

Menus

A dialogue is the series of exchanges which take place between a user and a computer system. Very often the dialogue commences by interaction with a series of menus which enable the user to choose an appropriate function. Additionally, many systems incorporate a hierarchical menu structure. Figure 6.7 shows three screens, all of which are menus. Note that they exist in a simple hierarchy.

Menus are represented diagrammatically by a hexagon. Figure 6.8 is the diagrammatic specification of Figure 6.7.

Note a special oval icon annotated QUIT is used to specify an exit from the system.

Functions or transactions

A function provides the processing for a transaction. Functions are represented by an ellipse. Figure 6.8 requires functions for invoice

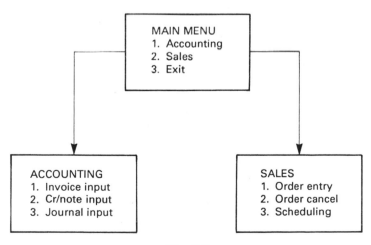

Fig. 6.7

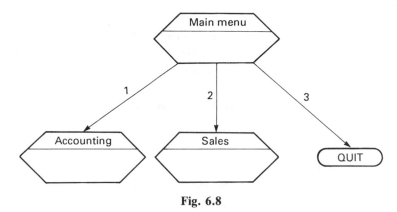

Fig. 6.8

input, credit note input, journal input, order entry and order cancel. 'Scheduling' calls another menu in the hierarchy. The specification of the system depicted by Figure 6.7 is enhanced from Figure 6.8 to show the functions and further menus by Figure 6.9.

Branching

So far the specification does not indicate which process or menu should follow the completion of a function. The oval is used to indicate a branch. Figure 6.10 adds the branches to the earlier specification to show a dialogue specification. Note from each menu or function the ESCape key takes the system *back* to the previous level.

A shorthand method exists for specifying this control. This is shown in Figure 6.11. Whenever the ESCape key is pressed the system reverts back a level. This facility can also be used to specify a global QUIT shown in Figure 6.12. PF10 depressed at any stage exits the system, while PF1 calls a help facility shown in Figure 6.13. GLEX is a branch back to where PF1 was invoked.

There are examples where the system should restrict this global facility. The example shown in Figure 6.14 shows a menu where the product price table is available via key PF5 to all menus and functions, with the exception of the function, stock processing. By using a named arrow with 'off', specific globals may be turned off. Note that the globals are turned *on* when the user returns to the start menu.

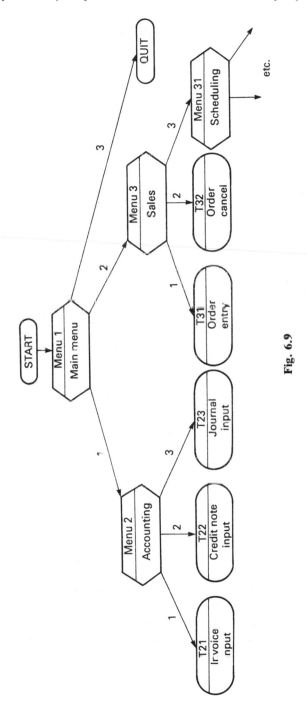

Fig. 6.9

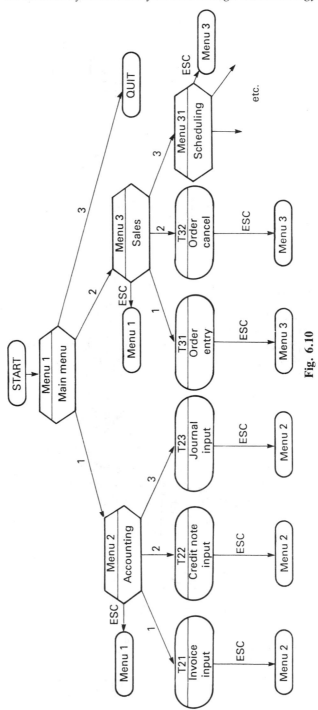

Fig. 6.10

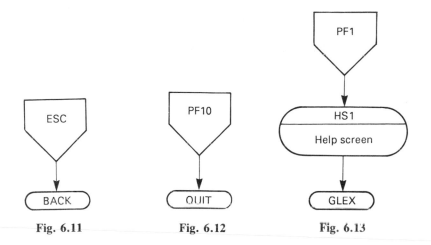

Fig. 6.11   Fig. 6.12   Fig. 6.13

Dialogue development from data flow diagrams

The data flow diagrams can be used as an aid to the development of the dialogue specification. A level 1 DFD will generally comprise a number of high level functions. This DFD should be mapped into a new menu, with each DFD function being a selection from the menu which would call a lower level menu representing a level 2 DFD.

The level 2 could comprise functions which are themselves further decomposed, and primitive functions. Each primitive function should be mapped to a dialogue function, the others to further menus.

This method will yield the basic dialogue hierarchy. The detailed control method needs to be considered, together with the requirements for data look-up and privacy. Special consideration should be given to appropriate links across the hierarchy.

Control is specified using control keys, branches and globals. Privacy is again specified by control keys and by turning off globals. More sophisticated privacy can be specified using a start up function which must execute before the initial menu is displayed. In this case a final dialogue icon is useful, the condition icon. Figure 6.15 shows such a use of this icon. It provides a conditional branch feature.

Dialogue prototyping

A dialogue prototyping should be produced to visualise for the user the menu structure and the user interaction with the menu structure. The

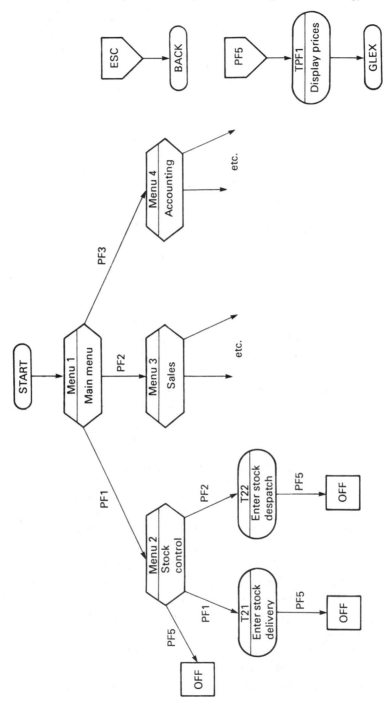

Fig. 6.14

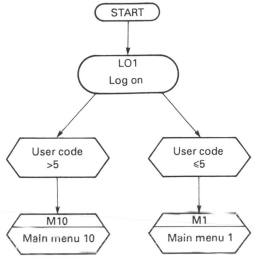

Fig. 6.15

user will be able to understand, comment on and agree the manner in which system functions are invoked.

A dialogue prototype may be built using the same software products used for input and output prototyping. This will enable the linking of the dialogue prototype and the input output prototype to visualise the total user interface to the system.

6.2.5 Task 2.10: Document the function descriptions

A function description is required for every primitive function on the required DFDs. The number of levels of decomposition produced should have ensured that all primitive functions can be documented easily, concisely, completely and unambiguously.

The input and output for each function will have been documented by tasks 2.6 and 2.7; it is the detailed processing that is now required to be documented. Many techniques may be used to document the processing including pseudo-code, structured English, decision tables and decision trees. These are techniques well known to the analyst, examples of each are given in Figure 6.16. They all provide the detailed documentation of the function also shown in Figure 6.16.

The function descriptions must be in business terms. It will be the user

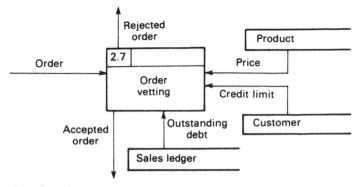

Pseudo-code
Compute order value: = order qty. × price each
Obtain credit limit
Obtain outstanding debt
Compute credit remaining: = credit limit – (order value + outstanding debt)

if credit remaining ≥ 0
  then accept order
  else if special clearance obtained
    then accept order
    else reject order.

Structured English

IF order value + outstanding debt less than credit limit
  THEN accept order
    ELSE IF special clearance obtained
      THEN accept order
      ELSE reject order.

Decision tree
Order value + outstanding debt less than credit limit?

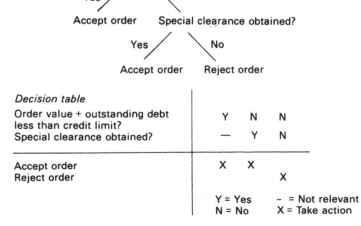

Decision table			
Order value + outstanding debt less than credit limit?	Y	N	N
Special clearance obtained?	—	Y	N
Accept order	X	X	
Reject order			X

Y = Yes   – = Not relevant
N = No   X = Take action

Fig. 6.16

who will agree the specification, and the user will generally only understand function descriptions which are written in business terms. The methods used for the function descriptions must suit the user. If the user presented much of the current processing using tables, then decision trees and tables may be appropriate. If the user has experience of computing systems and algorithm specification, then pseudo-code might be appropriate.

Stage A produces a business specification of requirements. The user must review and agree the specification; it must, therefore, be in a language acceptable to and understandable by the user.

6.3 VALIDATION OF THE SPECIFICATION

6.3.1 Task 2.11: Create the entity function matrix

The entity function matrix provides a third view of the system, a dynamic view. It charts the effect, in time, of functions on entities. Data flow diagrams and entity models do not provide a view which involves time. They provide a view of the data flow and of the data structure with no reference to time. The entity function matrix and the entity life histories, task 2.12, include the effect of time. They document events in time and their effect on the data. The sequence of events is important, not their real timing.

This third view of the system allows questions such as the following to be answered:
- Does each entity have a function which inserts it and a function which deletes or archives it?
- Is each entity accessed after being inserted?
- Which entities are referenced by a function?
  The major questions to be answered are:
- Does each entity possess a complete life?
- Does each function behave properly?

Functions form the columns of the entity function matrix, entities the rows. The matrix outline is, therefore, created by listing all the functions and entities within the system.

The required physical DFDs provide the source of the function list. The retention of the function's DFD number on the matrix provides a useful cross reference. The entity model from task 2.2 provides the list

ENTITY/FUNCTION MATRIX

SYSTEM:	EXAMPLE	DATE:
AUTHOR:	G. CUTTS	PAGE: 1 of 1

Entity name \ Function name	1 Order entry	2 Produce order/ deliv./inv. report	3 Stock check	4 Produce invoice	5 Record delivery success							
Customer												
Invoice												
Invoice line												
Order head												
Order line												
Price												
Product												
Delivery												

Fig. 6.17

of entities. Figure 6.17 shows the skeleton matrix for the system documented in Figures 5.5 and 5.8, the required physical DFD and the entity model, respectively.

The entries within the matrix may contain one or more of the letters

ENTITY/FUNCTION MATRIX

SYSTEM: *EXAMPLE*	DATE:
AUTHOR: G. CUTTS	PAGE: I of I

Function name / Entity name	1 Order entity	2 Produce order/ deliv./inv. report	3 Stock check	4 Product invoice	5 Record delivery success								
Customer	R	R		R									
Invoice		R		I									
Invoice line				I									
Order head	I	R		M									
Order line	I	R	M	M									
Price				R									
Product			M	R									
Delivery		R	I	M	M								

Fig. 6.18

I (insert), M (modify), R (read), D (delete), or A (archive) or space for 'no effect' on the entity.

Figure 6.18 shows the entity function matrix for the system shown in Figures 5.5 and 5.8.

The entries are made by reference to the DFDs. All data flows between functions and data stores result in one or more entries into the matrix.

The following paragraph refers to Figures 5.5 and 6.18:

Function 1, order entry, reads the entity, customer, since the arrow points from the data store to the function. Function 1 also inserts the entities, order head and order line, since the arrow points to the data store. A double pointed arrow indicates reading and writing, a modification of the entity. The modification may be a genuine modification or one with a null write representing a deletion or archive of the entity occurrence.

Some functions may perform more complicated processing; entries such as I/M are possible.

Entity life

Each row of the matrix shows the life of an entity. It shows which functions insert the entity, which functions read or modify it and which functions delete or archive it. Each row should comprise at least one I (insert), one R (read) or M (modify) and one D (delete) or A (archive).

Entities such as customer and supplier exist in a system so that orders, deliveries and invoices can be inserted, processed and deleted. Very often the function which inserts the entity customer or supplier is overlooked during investigation and analysis. It is an exception as opposed to operational processing. There will, therefore, be no documented insert function on the DFDs. The entity row in the matrix will comprise reads, R, and modifies, M, but no I, insert. Such omissions need to be corrected.

Similarly, the function on the current system which removes customer entities out of the physical data store may well not have been documented. Indeed, such a function may never take place, the data store being allowed to grow indefinitely. This situation will be shown up in the entity function matrix by the absence of a D, delete, within an entity row.

If a row comprises a single I and a single D or A, then questions need to be asked regarding the need for the entity. Is it inserted, never accessed and then deleted or archived?

Multiple occurrences of I, D or A also require further investigation. If an entity can be inserted by more than one function, there must be a

Product	1.3 Create new product entity	1.4 Record delivery into stock	2.1 Record delivery to customer	4.3 Discontinue product
Product	I	I/M	M	D

Fig. 6.19

distinct possibility of inconsistency between the two functions. Figure 6.19 shows an entity row with two insert functions.

Function 1.3, create new product entity, results from a management decision to stock and sell a new product. When deliveries of stock arrive they are recorded in the system by function 1.4 which modifies the quantity in stock. If a delivery of stock is made for which no product entity exists, then function 1.4 automatically inserts the entity. This prevents the transaction being rejected, saving time and effort. However, it also creates an entity occurrence for every wrongly coded delivery note, resulting in a total loss of control of the database. Function 1.4 should be allowed only to modify existing occurrences, rejecting transactions for which no entity occurrence exists. Function 1.4 should therefore be allowed modification status but not insert status.

Multiple delete or archive entries are also possible. The provision of multiple functions which can delete or archive an entity occurrence leads to auditing problems. Multiple deletes, archives and inserts require careful investigation with subsequent reduction to single entry only, if possible.

The entity lives shown in Figure 6.18 need careful investigation. Three new maintenance functions are required, function 6, customer maintenance, function 7, product maintenance, and function 8, price table maintenance. Each of these functions will insert, modify and delete entity occurrences.

ENTITY/FUNCTION MATRIX

SYSTEM: EXAMPLE	DATE:
AUTHOR: G. CUTTS	PAGE: 1 of 1

Entity name \ Function name	1 Order entry	2 Produce order/ deliv./inv. report	3 Stock check	4 Produce invoice	5 Record delivery success		6 Customer maintenance	7 Product maintenance	8 Price tables maintenance		
Customer	R	R		R			I/M /D				
Invoice		R		I							
Invoice line				I							
Order head	I	R		M/A							
Order line	I	R	M	M/A							
Price				R					I/M /D		
Product			M	R				I/M /D			
Delivery		R	I	M/A	M						

Fig. 6.20

The entities invoice and invoice line are inserted and never accessed. This is satisfactory in the system, since the entities are further processed by a sales accounting system. An invoice entity occurrence is eventually created for order head, order line and delivery entity occurrences. Function 4 will be allowed to archive the entities order head, order line and delivery subject to an invoice having been inserted into the database. Each entity now possesses a complete life.

Figure 6.20 shows a revised entity function matrix. Three new functions have emerged and one has been amended. These new functions need to be added on to the required DFDs to complete the documentation.

6.3.2 Task 2.12: Create the entity life histories

Each row of the entity function matrix provided a list of functions with each function's effect on an entity. The entity function matrix cannot show, for every entity, the sequence of functions, nor can it show the effect of abnormal events. Each abnormal event requires a function to process the event.

Diagram conventions

There are several diagram conventions for entity life histories. The convention illustrated first is based on the theory of Petri nets. Four structures are used for entity life histories when based on Petri nets. The structures are shown in Figure 6.21.

The oval is used as a diagrammatic start and end symbol. It is annotated with the entity name.

The square is used for a function. The function name is written into the square and the function's DFD number is written into the box in the bottom right hand corner of the square. Its effect on the entity is written into the bottom left hand box. The first function after the start oval must be the insert of the entity into the database.

The circle is used to show the status of the entity. Circles are numbered and may contain a description of the status, e.g. order despatched.

Arrows indicate the transition from one status to another by the action of a function. Figure 6.22 shows an entity life history. The entity order is inserted into the database by function 1.1, order entry. The

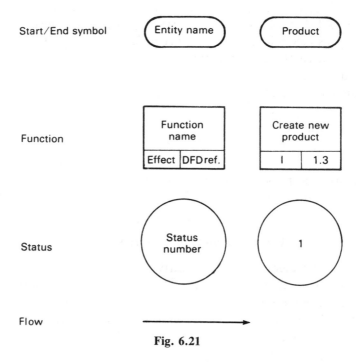

Fig. 6.21

order then takes on the status 1. It is modified by functions 3.2, produce delivery note, and 3.3, invoice, then deleted by function 4.7, receive payment of invoice. Note that function 3.2 cannot take place unless the status of the order is 1, that is, an order exists. After function 3.2, the status is set to 2. Similarly, function 3.3, invoice, has a 'valid previous' status of 2 and a 'set to' status of 3, and function 4.7, receive payment of invoice, has a 'valid previous' status of 2 and no 'set to' status.

Functions which insert, modify, delete or archive entities are shown on the entity life history. All these functions modify the entity status; functions which only read do not modify the status.

Simple life

Many entities have simple lives. An occurrence of an entity is inserted, read, perhaps many times, and eventually deleted. A simple life is shown in Figure 6.23. Note that since read functions do not affect an entity's status, they are not shown on the entity life history.

A minor extension of this simple life is shown in Figure 6.22. The

ENTITY LIFE HISTORY

SYSTEM: EXAMPLE LIFE	DATE:
AUTHOR: G. CUTTS	PAGE: 1 of 1

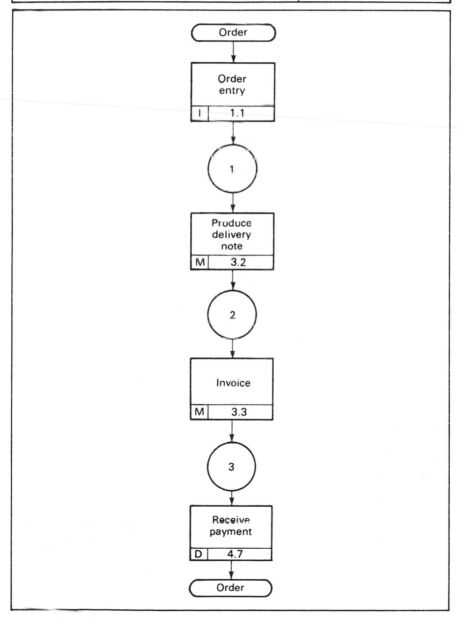

Fig. 6.22

ENTITY LIFE HISTORY

SYSTEM: *SIMPLE LIFE*	DATE:
AUTHOR: *G. CUTTS*	PAGE: of

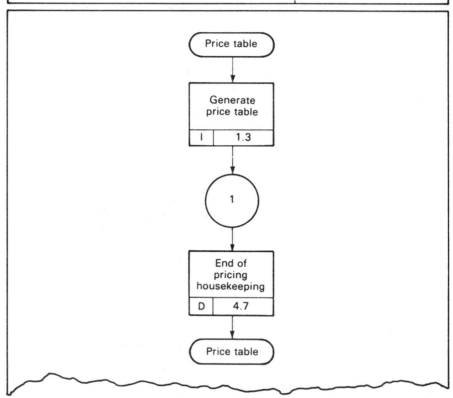

Fig. 6.23

entity order is inserted, modified to show the quantity delivered by the function, produce delivery note, further modified to indicate that an invoice has been produced by the function, invoice, and finally deleted by the function, receive payment.

Only one delivery is allowed against each order if the entity order has the life history shown in Figure 6.22. Multiple deliveries are allowed in Figure 6.24.

ENTITY LIFE HISTORY

SYSTEM: MULTIPLE DELIVERIES	DATE:
AUTHOR: G. CUTTS	PAGE: 1 of 1

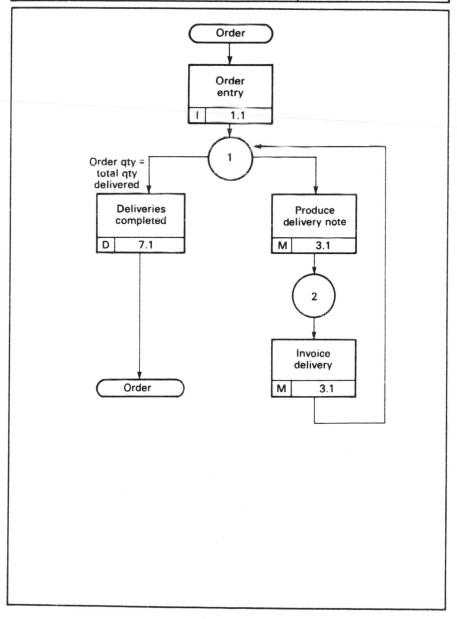

Fig. 6.24

ENTITY LIFE HISTORY

SYSTEM: SEPARATE 1st & last DELIVERIES

DATE:

AUTHOR: G. CUTTS

PAGE: 1 of 1

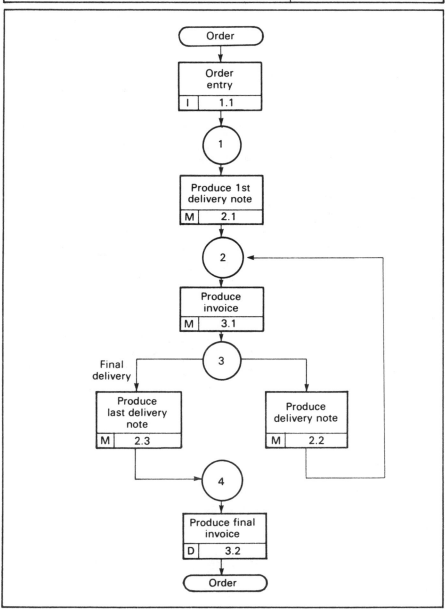

Fig. 6.25

The function, produce invoice for delivery, resets the status to 1, which allows the function, produce delivery note, to execute. However, when the status of the order entity is set to 1, two functions can execute, produce delivery note, and deliveries completed. The choice of function to execute can only be made based on further information, for example the total quantity delivered to date. The entity life history can indicate choice by simply annotation of the arcs. Note also that the order cannot be deleted if a delivery note exists which has not been invoiced, since the status would be 2 which would prevent the function, deliveries completed, from executing.

Figure 6.25 shows an entity life history for the entity, order, where there exists a requirement to show separately the first and last deliveries.

Function 2.1, produce first delivery note, is very slightly different to functions 2.2, produce delivery note, and 2.3, produce last delivery note. They differ only in wording: 'this is your first delivery, an interim delivery, your last delivery'. The status indicators now provide more information concerning the entity:

Status 1 order accepted, no deliveries made
          2 order accepted, 1 or more deliveries made, last not invoiced
          3 order accepted, 1 or more deliveries made, and invoiced
          4 order accepted, all deliveries made, last not invoiced

Diagram structures

Three structures have been introduced: sequence, selection, and iteration. Figure 6.26 shows these simple constructs. They equate to standard programming structures and provide an excellent relationship with process design. Data design and process design possess a common approach. The structures also allow entity life histories to be developed top down, in a proven, well-understood fashion.

Concurrent functions

Sequence, selection and iteration only allow functions to execute in sequence, that is one function followed by another. There are many situations where sequence is not important in that functions may execute concurrently.

The entity, customer account, may be modified to record deliveries

Sequence

Selection

Iteration

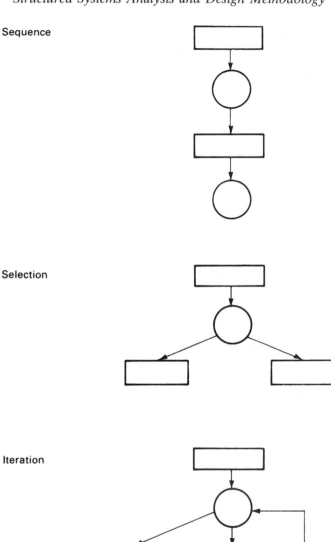

Fig. 6.26

ENTITY LIFE HISTORY

SYSTEM: EXAMPLE	DATE:
AUTHOR: B. CUTTS	PAGE: 1 of 1

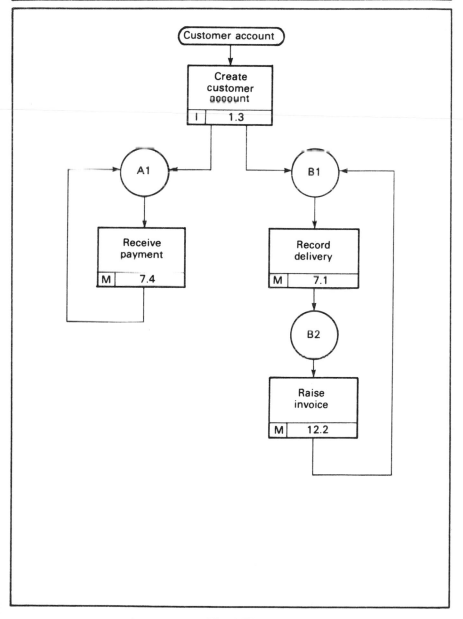

Fig. 6.27

ENTITY LIFE HISTORY

SYSTEM: EXAMPLE	DATE:
AUTHOR: G. CUTTS	PAGE: 1 of 1

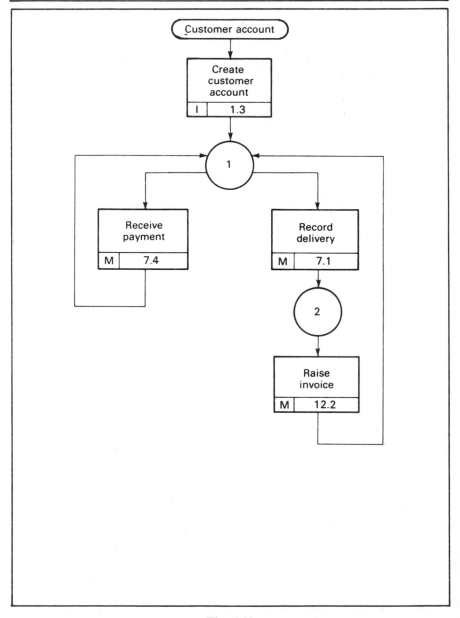

Fig. 6.28

and invoices raised against the account concurrently, with modification to record the total value of payments received. This entity life history is shown in Figure 6.27.

The easiest way of handling the status indicators is to create a separate one for each of the concurrent lives. It is important to understand that this entity life is different from that shown in Figure 6.28, since, with the concurrent lives, payments may be received when the B status is set to 2. This is not possible in Figure 6.28.

The introduction of a second status indicator creates a status vector, a generalisation of the simple status indicator. The status vector (Figure 6.27) has the following interpretation:

    A B
Status vector <1 1> account created, functions 7.1 and 7.4 may
                       execute concurrently
            <1 2> account created, functions 12.2 and 7.4 may
                       execute concurrently.

It could be argued that no payment should be received until the first invoice has been issued; this could be easily incorporated into the entity life history.

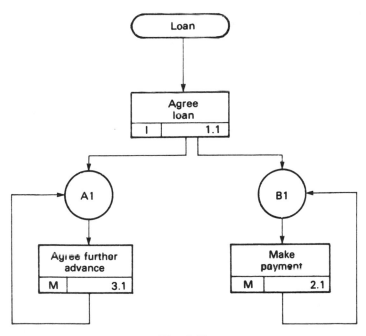

Fig. 6.29

ENTITY LIFE HISTORY

SYSTEM: ExAMPLE	DATE:
AUTHOR: G. CUTTS	PAGE: I of I

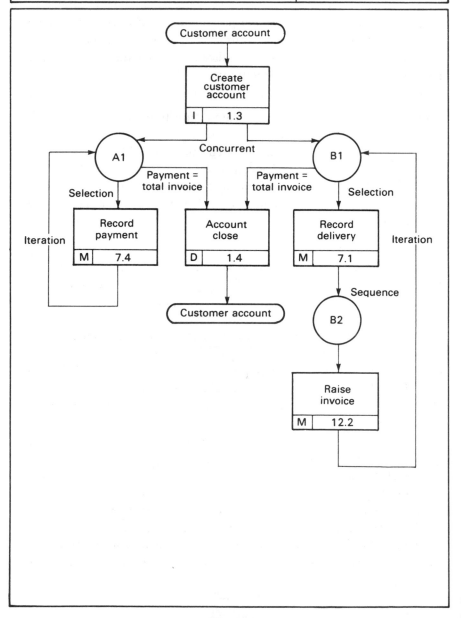

Fig. 6.30

Entities may, therefore, have concurrent lives, represented by the diagram construct shown in Figure 6.29.

Each individually executing entity life may use sequence, selection and iteration constructs and, of course, may be developed using top down design techniques.

The conclusion of currently executing lives does not require any new construct. The selection construct with identical criteria applied to all lives is sufficient. All lives must provide an input to the function. Where more than two lives exist, concurrency may be gradually reduced. Figure 6.30 shows a possible life history for the entity, customer account.

A choice of functions can take place when, and only when, the status vector is <1 1>. Either functions 7.1 and 7.4 may execute singly or concurrently, or, if the total of payments received equals the total invoice value, then function 1.4 may execute. The choice does not mean that function 1.4 must execute when the criterion is satisfied, only that it may execute.

Abnormal lives

The documentation of standard lives for each entity is followed by careful analysis of each life for abnormal but possible situations. For example, 'what happens to customer accounts where payment is never received?'. In most cases, arcs are required to allow premature deletion or archiving of an entity due to a variety of reasons.

The status points to be linked to the deletion activity require careful choice, as not all status points should be linked. For example, if a customer account is being closed because of bad debts, all deliveries should still be invoiced for accounting purposes before the account is closed. Status <1 2>, Figure 6.27, should not, therefore, be linked to the account closing function.

All choice arcs should be annotated. These abnormal life arcs require choice between continuing with the normal life or executing some abnormal termination. They should be annotated with the choice criterion. Figure 6.31 shows an abnormal termination arc.

The account can only be closed when the status is 3. This may be reached from status 1, when there is no debt, via a normal closure route, or from status 2, when there are payments outstanding. In this case, status 3 is only reached via the function, 12.2, claim insurance. The

ENTITY LIFE HISTORY

SYSTEM: ABNORMAL LIFE	DATE:
AUTHOR: G. CUTTS	PAGE: 1 of 1

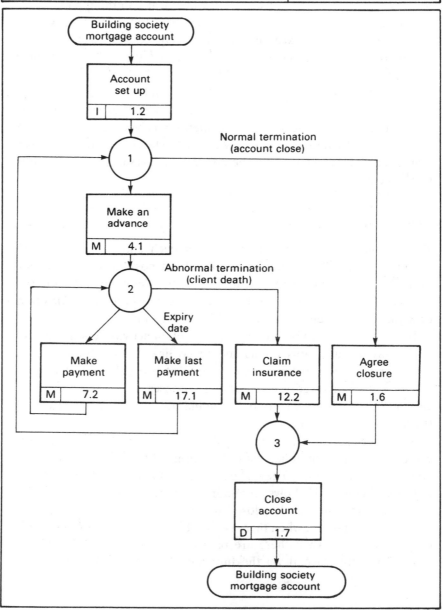

Fig. 6.31

function, claim insurance, would probably have been identified during stage A, analysis, and would be recorded on the DFDs. This is not always the case with abnormal terminations, resulting in the identification of new functions. These new functions should be recorded on the required DFDs, to keep the documentation up to date.

Complete life

The construction of an entity life history ensures that each entity possesses a complete life. Each occurrence of the entity in the database must be at some appropriate stage in its life, the stage being represented by the status indicator or status vector.

Moreover, each complete life possesses a beginning, a normal life, a possible abnormal life, and a series of functions which always lead to an end to the life. Entity occurrences are inserted, behave properly and are eventually deleted or archived. The designer of the database can be assured that the functions specified by the system act upon entities within the database in a prescribed and controlled fashion.

The inclusion of the status as an attribute of the entity provides some of the control. Each entity occurrence will have associated with that occurrence the status. This instantly limits the number of valid processes which may have an effect on the entity. Each process specification will first test the entity status before execution, setting the new status after successful completion of processing. Entity status is important in part 4, logical process design.

Petri net convention

The entity life history convention using Petri net theory enables the dynamic concept of an entity life history to be modelled by the movement of tokens.

A Petri net is made up of places – the circles or states on the entity life histories – and transitions – the functions of the entity life histories. A marked Petri net includes the idea of tokens which reside on the places. The reader should consult Wolfgang Reisig (1985) for further information.

Figure 6.32 shows a Petri net; Figure 6.33 shows a marking of the Petri net. The initial marking, Figure 6.33, shows a token on place 'a', with zero tokens on all the other places. Using Petri net terminology,

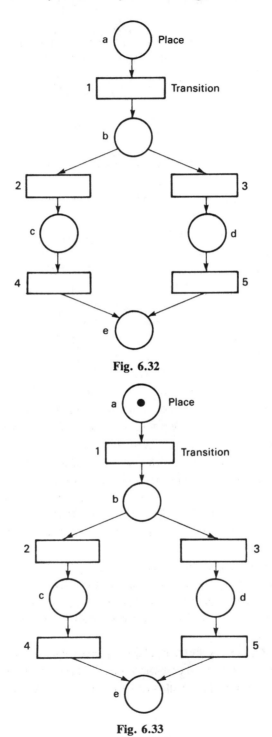

Fig. 6.32

Fig. 6.33

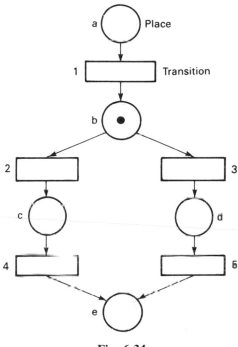

Fig. 6.34

transition 1 may now occur; it possesses a token on all of its input places. A token on place 'a' represents a state of the entity life history where the only valid function which may execute is function 1.

A successful occurrence of transition 1, that is successful execution of function 1, removes the token from place 'a' and deposits it on place 'b'. The 'valid previous' and 'set to' status for each function within the entity life history is represented by the position and movement of the tokens. Figure 6.34 shows the marking or status after function 1 has successfully executed.

Function 2 or 3 may now execute as both have a token on their input places. Note that this construct models choice or if. . . then. . . Either functions 2 and 4 execute or functions 3 and 5 execute. Figure 6.35 shows the movement of tokens representing execution of functions 2 and 4.

When a transition occurs or a function executes, a token is deposited on all its output places. Petri nets may, therefore, be used to model parallel execution as shown in Figure 6.36.

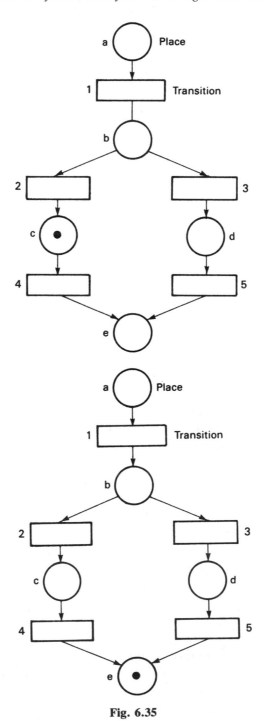

Fig. 6.35

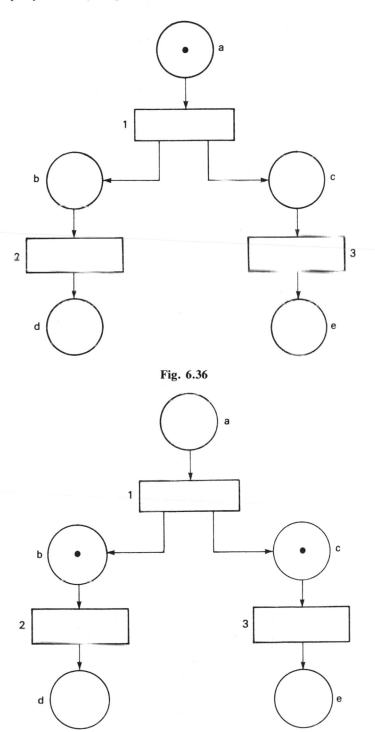

Fig. 6.36

Fig. 6.37

The successful execution of function 1 removes the token from place 'a' and deposits tokens on to *all* the output places, 'b' and 'c' shown in Figure 6.37. Functions 2 and 3 are now enabled and may execute in parallel. The status is given for each entity life history by the position of the tokens. Essentially this is a vector with one entry for each place. The marking or status of Figure 6.36 is:

1
0
0
0
0

The marking or status of Figure 6.37 is

0
1
1
0
0

This mathematical representation of the marking may be enhanced by a representation of the Petri net. Two matrices can represent a Petri net,

Function	1	2	3
Place			
a	I	0	0
b	0	I	0
c	0	0	I
d	0	0	0
e	0	0	0

Fig. 6.38 Pre-condition matrix

Function	1	2	3
Place			
a	0	0	0
b	I	0	0
c	I	0	0
d	0	I	0
e	0	0	I

Fig. 6.39 Post-condition matrix

a pre-condition matrix (Figure 6.38) and a post-condition matrix (Figure 6.39).

Read each column of the matrix as follows. For function 1 to execute, there must be a token on place 'a', the 'valid previous' status for function 1, obtained from the pre-condition matrix. After successful completion of function 1, tokens must be deposited on places 'b' and 'c', the 'set to' status for function 1, obtained from the post-condition matrix.

This mathematical format for entity life histories provides a convenient method of recording Petri nets and thus entity life histories in a computer format. Sophisticated analysis of entity life histories is now possible, which is outside the scope of this book. Current research on Petri nets may be found in *The Advances in Petri Nets* (Ed. G. Rozenberg) and in the proceedings of the annual International Conference on the Theory and Application of Petri Nets (published in the Advances).

Hierarchical convention

Other conventions for entity life histories are based on hierarchical and network constructs. Figure 6.40 shows an entity life history using the hierarchical constructs. It is equivalent to Figure 6.22. The 'valid previous' and 'set to' status for each function are shown either side of a solidus, e.g. 2/3, where the 'valid previous' status is 2 and the 'set to' status is 3.

The asterisk notation is used to indicate iteration. Figure 6.41 is equivalent to Figure 6.24 where iterations of functions 2.1 and 3.1 are allowed.

The constructs in Figures 6.40 and 6.41 show sequence and iteration. Selection is achieved using the circle notation to represent an optional function. An optional null function is also required. Figure 6.42 shows an order being inserted either via order entry or via a function which generates orders to satisfy a previously negotiated contract.

Abnormal lives are shown using selection and a notation using Q and R for quit and resume. Figure 6.43 shows a possible order cancellation. The entity life history may quit at the functions indicated and resume with the cancellation of the order. Alternatively, if the order is processed normally, then the possible cancellation null function should be chosen.

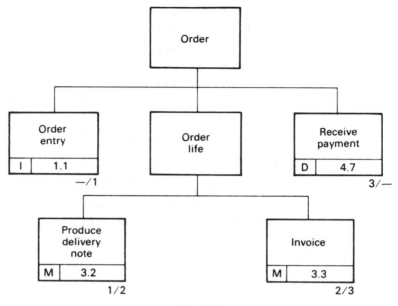

Fig. 6.40

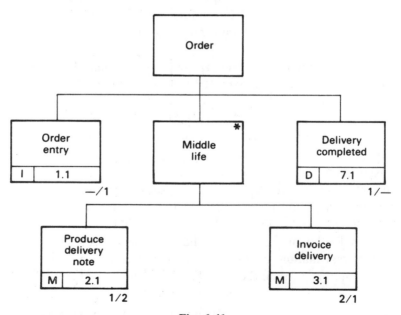

Fig. 6.41

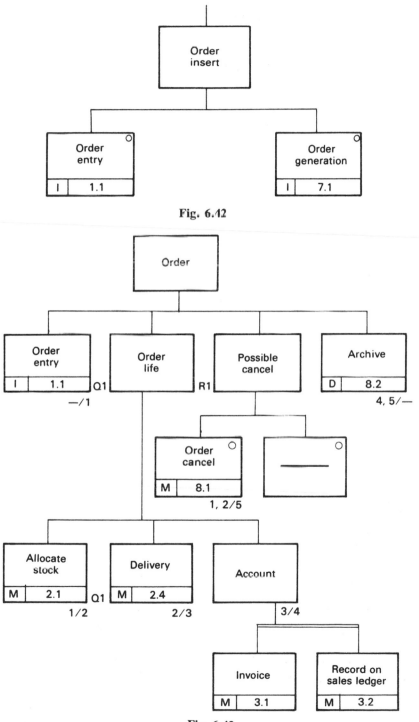

Fig. 6.42

Fig. 6.43

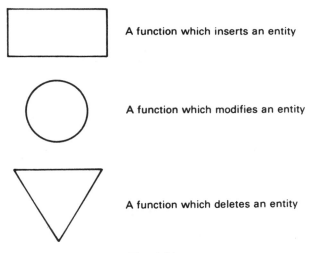

A function which inserts an entity

A function which modifies an entity

A function which deletes an entity

Fig. 6.44

The final construct required is also shown in Figure 6.43. Parallel horizontal lines indicate that functions 3.1 and 3.2 may execute in any sequence or in parallel.

Network convention

Network entity life histories are based on the diagram conventions shown in Figure 6.44. The entity life histories equivalent to Figures 6.22 and 6.24 are shown in Figures 6.45 and 6.46.

The status indicator is dealt with in exactly the same fashion as the hierarchical entity life histories. The asterisk is used either within a box (Figure 6.46) or within an individual modify circle to represent zero or many iterations of the box or circle.

Selection is achieved by simply following the arcs on the network (Figure 6.47) and parallel functions are shown using the box notation (Figure 6.48). Abnormal lives are easily added by using the arcs in a similar fashion to the arcs used by the Petri net approach.

6.4 PART 2 SUMMARY

The specification of business requirements comprises twelve tasks which may be divided into four sections.

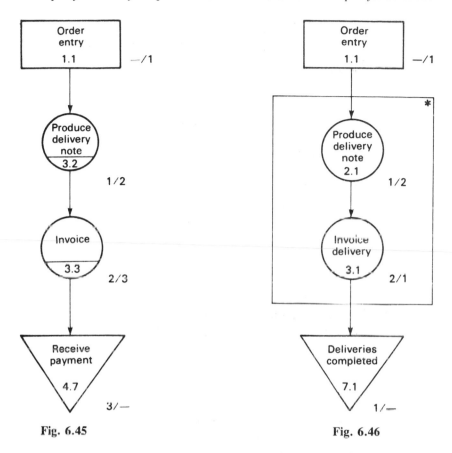

Fig. 6.45   Fig. 6.46

Section 1, tasks 2.1 and 2.2 Develops the required logical model. The outputs from this initial section are the required logical DFDs and a required entity model. Solutions to problems with the current system and requirements for the new system should be incorporated into the required logical models.

Section 2, tasks 2.3 to 2.5 Creates the outline business specification by examining different business options and by selection of a particular option for further development.

Section 3, tasks 2.6 to 2.10 Provides the detailed documentation for the business specification.

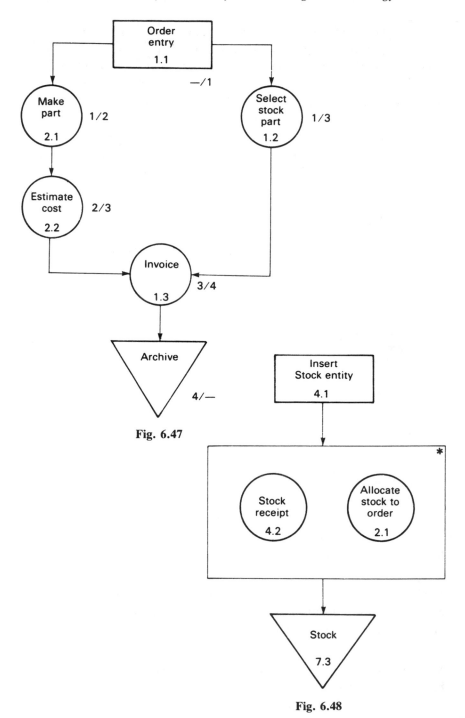

Fig. 6.47

Fig. 6.48

Section 4, tasks 2.11 and 2.12 Creates the entity function matrix and the entity life histories, to provide a third view of the system, a dynamic view, showing the effect of functions on entities. The entity life histories can then be analysed to determine if additional functions, or modifications to functions, are necessary in order to ensure that the entity has a complete life in all circumstances.

Any additional functions or modifications to functions must be documented on the required DFDs. It may be necessary to return to previous tasks to bring the documentation up to date.

Chapter 7
Logical Data Design

7.1 INTRODUCTION

This chapter describes the third part of structured systems analysis and design, shown in Figure 7.1. Part 3 uses the required physical DFDs, the entity model, the entity descriptions and the input and output descriptions to produce a logical entity model and a revised set of entity descriptions. The logical entity model will form the basis for the database design; and the revised entity descriptions, together with the data dictionary, will form the basis for the record descriptions.

7.2 TASK 3.1: SELECT DATA STRUCTURES

The normalisation task, while relatively straightforward, can be time consuming. Careful selection of data structures as candidates for normalisation is therefore necessary. Sufficient data structures need to be analysed to ensure the resultant entity model is complete. Normalisation should be considered as a further process in the refinement of the entity model, not a technique to produce the final version. All the techniques used in the construction of the final entity model can be used to further refine the model. If additional data structures are created during the later stages of the project, then normalisation can be performed on the new data structures resulting in yet further refinement of the entity model.

It is important to identify the major data structures; these will include:
(1) output documents/screen content
(2) input documents/screen content.

In many cases it is only necessary to analyse the output documents and screen content as these represent the data necessary to be stored by the system. A list of data structures can be easily obtained from the level 1 required physical DFD, the system's input and output. It is

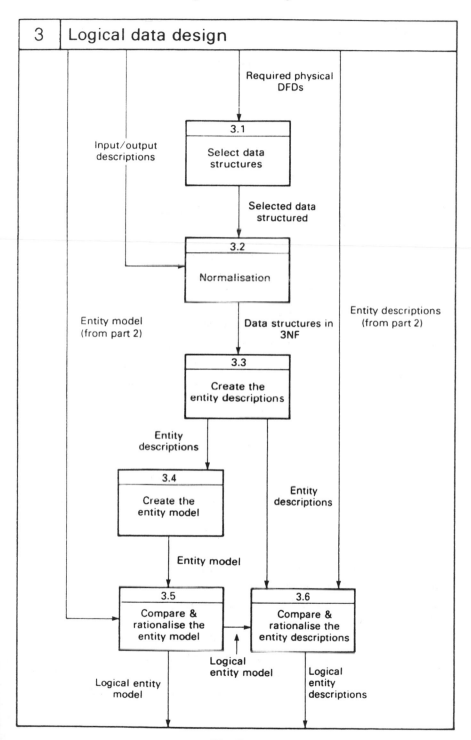

Fig. 7.1

Order	Customer number
	Customer order number
	Customer address
	International Standard Book Number
	Title
	Author
	Price each
	Quantity ordered

Fig. 7.2

| New customer | Customer number |
| | Customer address |

Fig. 7.3

Textbook	International Standard Book Number
	Title
	Author
	Price each

Fig. 7.4

particularly important to include output which was not in the original system. Entities and relationships necessary to generate new output may not have been included in the entity model which was constructed as part of systems analysis, since this was a development of the data structures supporting the current system.

The first sub-task of the normalisation task involves listing all attributes of a data structure. Figure 7.2 gives a list of attributes of a sales order document for text books, and Figure 7.3 gives a list of attributes of a new customer agreement form. Note that the attributes in Figure 7.3 are a subset of the attributes in Figure 7.2, so there is nothing to be gained by applying the normalisation process to the new customer form. Similarly, the test book data structure shown in Figure 7.4 may be eliminated from the normalisation task.

Task 3.1 yields a set of data structures as candidates for normalisation.

7.3 TASK 3.2: NORMALISATION

Normalisation consists of four sub-tasks. It is a technique for transforming complex data structures into simple tables which form the basis of the third normal form entity model.

Normalisation sub-tasks

- Convert the unnormalised data structures into first normal form (1NF).
- Convert each of the data structures generated in 1NF into second normal form (2NF).
- Convert each of the 2NF data structures into third normal form (3NF).
- Validate the 3NF data structures.
  Each sub-task is illustrated using examples.

Representation of data structures

Figure 7.5 shows, in table format, a data structure representing orders for fruit. The table contains three rows providing details of a given customer's order. Because of the problem of representing rows of varying length on a single line, it is more usual to present the table shown in Figure 7.6, which still comprises three rows.

A thorough understanding of the data is necessary before the process of normalisation can be successfully undertaken.

Hidden meaning

In Figure 7.6 the rows are organised in priority sequence. Apples and bananas will be supplied to Jones before Allen and finally Williams. This is very common with manually completed list, the orders to be supplied first are sorted to the head of the list. Normalisation considers only the logic of the data structure, not the physical presentation. To ensure the information contained in the physical presentation of the data is not lost, an extra column, priority, must be added. The sequence or relative positions of rows and columns must not conceal information about the data structures.

Customer number	Customer name	Product number	Product name	Qty.	Price	Product number	Product name	Qty.	Price
J27	Jones	A01	Apples	10	2.00	B14	Bananas	12	3.00
A12	Allen	B14	Bananas	50	3.00	P16	Pears	20	1.50
W01	Williams	A01	Apples	20	2.00				

Fig. 7.5

Customer number	Customer name	Product number	Product name	Qty.	Price
J27	Jones	A01	Apples	10	2.00
		B14	Bananas	12	3.00
A12	Allen	B14	Bananas	50	3.00
		P16	Pears	20	1.50
W01	Williams	A01	Apples	20	2.00

Fig. 7.6

Name	Date	Port	Port	Time
Spirit	14 July 87	Calais	Dover	14.00
Herald	14 July 87	Dover	Dunkirk	10.00

Fig. 7.7

Meaningless data names

Figure 7.7 gives a set of column headings, many of which are ambiguous. The headings must be unique and meaningful for a proper understanding of the data. Figure 7.8 represents the data structure after further discussion with the user. Each column is now uniquely defined.

7.3.1 First normal form (1NF)

The first sub-task of normalisation is to remove the groups of repeated data within an unnormalised data structure.

Vessel name	Date of sailing	Port from	Port to	Departure time (local)
Spirit	14 July 87	Calais	Dover	14.00
Herald	14 July 87	Dover	Dunkirk	10.00

Fig. 7.8

Unique keys must be selected for the complete data structure and for each group of repeated data. Selection of keys is necessary for the normalisation process; however, the process will work for any correctly chosen key, but it is sensible to choose one that will be usable as part of the final design.

Choosing a key

(1) Any key chosen must be unique, that is it must uniquely identify a row.
(2) If a choice of keys is possible, lengthy keys and alphanumeric keys should be avoided.

A unique key for the order data structure (Figure 7.6) could be the customer number. This assumes that customers have one order only outstanding at any point in time. In this case, a single attribute provides the key – a simple key. This is not always possible and in some cases a compound key has to be formed from two or more attributes.

If the restriction on outstanding orders was not possible, then a key formed by two attributes would be necessary. A candidate key in this case would be the customer number linked with the date of the order.

In Figure 7.9 it is necessary to choose a compound key formed by the course code and the subject. Both attributes are required to uniquely identify a row. A simple method of marking the key is by underlining. Having chosen a key, 1NF is created by removing all groups of repeating attributes. A group can be anything from one attribute to very many. Start by listing, in a single column, the column headings. Figure 7.10 provides a list of column headings for the data structures shown in Figure 7.6. Note that repeated groups of attributes are marked by a vertical bar and optional attributes are shown enclosed by round brackets ( ).

Course code	Subject	Lecturer ID.	Time allocated	Qualification	Number on course
Comp. Stud.	Systems	GC	90	HND	60
Comp. Stud.	Prog.	DS	100	HND	60
Comp. Stud.	Bus. stud.	WW	60	HND	60
Inf. Tech.	Electronics	KE	60	B.Sc.	50
Inf. Tech.	Systems	GC	90	B.Sc.	50
Inf. Tech.	Prog.	DS	90	B.Sc.	50

Fig. 7.9

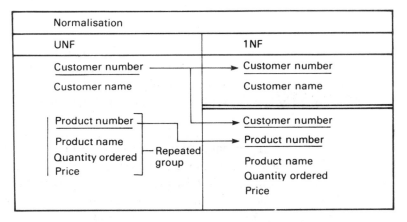

Fig. 7.10   **Fig. 7.11**

If a key has not already been chosen for the repeating group, it should now be chosen. In this case, product number seems an appropriate choice. First normal form is then created by performing the transformation shown by the arrows in Figure 7.11. Two data structures are created, the second structure having several occurrences in the database.

With a simple repeated group as in Figure 7.10, the conversion to 1NF is relatively easy. It is possible, however, to have nested repeating groups or more than one repeating group. The appropriate transformations are shown in Figures 7.12 and 7.13.

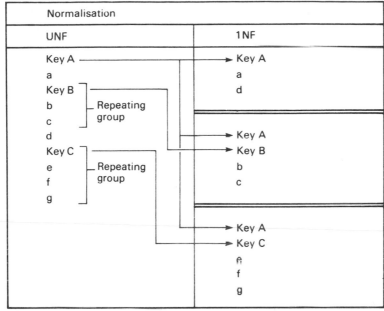

a to g are data items

Fig. 7.12

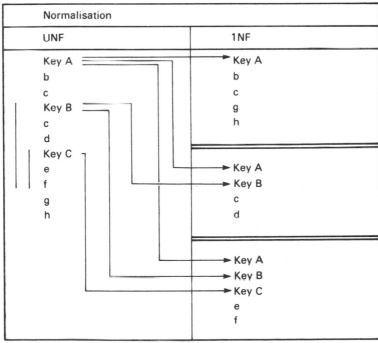

a to h are data items

Fig. 7.13

Normalisation	
1NF	2NF
<u>Customer number</u> Customer name	<u>Customer number</u> Customer name
<u>Customer number</u> <u>Part number</u> ◄────────┐ Part name ───────┘ Qty. ordered Price ──────────┘	<u>Customer number</u> <u>Part number</u> Qty. ordered
	<u>Part number</u> Part name Price

Fig. 7.14   Fig. 7.15

7.3.2 Second normal form (2NF)

The sub-task of moving from 1NF to 2NF only applies to data structures with compound keys. Data structures with a simple key in 1NF are automatically in 2NF.

The process requires that each attribute is examined to determine if it is dependent on the whole of the compound key or only part of the compound key. Attributes which only depend on part of the compound key are shown using arrows, Figure 7.14.

Part name and price only depend on the part itself, not on the customer number. The quantity ordered depends on the specific part on this order where we have identified it by the customer number.

Three data structures result and are shown in Figure 7.15. The attributes which only depend on the part number have been extracted to create a third data structure. This leaves only quantity ordered in the second data structure.

7.3.3 Third normal form (3NF)

The sub-task conversion from 2NF to 3NF is similar to the conversion from 1NF to 2NF. Instead of examining the relationship between non-key attributes and attributes in the key, the relationships between

Normalisation	
2NF	**3NF**
<u>Customer number</u> Customer name Address Hotel code ◄────────┐ Hotel address ────┤ Number of rooms ──────┘ Date of holiday Number of nights	<u>Customer number</u> Customer name Address Hotel code* Date of holiday Number of nights ――――――――― <u>Hotel code</u> Hotel address Number of rooms

Fig. 7.16   **Fig. 7.17**

pairs of non-key attributes and between pairs of key attributes are established.

Figure 7.16 shows a data structure in 2NF. (Customers only book one holiday at a time.) The arrows show the relationships between non-key attributes. Ask, 'Is attribute A dependent on attribute B or vice versa?'. This question must be repeated for all the non-key attribute pairs, and the key attribute pairs.

Extraction of the inter data dependency creates two data structures in 3NF, shown in Figure 7.17. The hotel address and number of rooms in the hotel do not change from holiday booking to holiday booking.

Hotel code is now a key to a new data structure. Since this attribute, hotel code, appears in the first data structure, it is marked by an asterisk and called a foreign key.

The choice of hotel number as the key for the new data structure was obvious and trivial. This is not always the case and careful examination of the real data may be necessary.

The data structure shown in Figure 7.18 is in 2NF, but not in 3NF, since a relationship exists between job title and salary scale. The 3NF data structures are either those shown in Figure 7.19 or those shown in Figure 7.20.

The key to the second data structure, salary scale/job title, needs to be established. A diagram (Figure 7.21) helps to establish which attribute should be the key. For each job title there exists only one

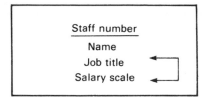

Fig. 7.18

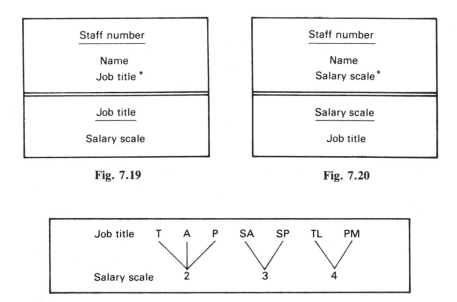

Fig. 7.19   Fig. 7.20

Fig. 7.21

salary scale, but for each salary scale a range of job titles is possible. Job title becomes the key and Figure 7.19 the 3NF representation. Alternatively, if the diagram in Figure 7.22, had resulted, then Figure 7.20 would be the correct 3NF representation.

7.3.4 Validation of the 3NF data structures

The normalisation task is based on sound mathematical principles; however, knowledge of the data and the application of common sense contributes good practice.

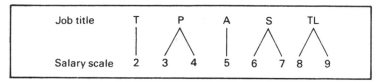

Fig. 7.22

Order number
Product code
Discount code

Fig. 7.23

It is good practice to validate, in detail, each 3NF data structure to eradicate errors in processing and to discover obscure dependencies.

Errors in the normalisation task can be established by asking the questions: 'For each unique value of the total key can each attribute possess one, and only one value?' and 'Is the value of each attribute directly dependent on the key or keys?'. For example, the value for discount code, Figure 7.23, can possess only one value for a particular order and product. The answer to the first question is 'Yes'.

However, by further investigation it can be established that the discount code depends on the customer identified by customer code. All orders for all products for any given customer have the same discount code. This data structure is, therefore, replaced by that shown in Figure 7.24, assuming that a data structure already exists linking customer code and order numbers. The answer to the second question is 'No'.

The section describing first normal form required that a key be chosen for each of the data structures. This key could be either a simple key or one made up from several simple keys, a compound key. The selection of a key is not always straightforward and often it is necessary to use a composite key or a generated key. A further problem sometimes exists where the key does not exist as a data attribute but is implied by the existence of some other attribute or attributes. Composite keys, generated keys and implied keys must be considered during normalisation.

Customer code
Discount code

Fig. 7.24

Equipment code	Description	Identification numbers	Date required
D	Desk	1–100	1/10/80
D	Desk	200–300	1/10/81
C	Chair	1–100	1/10/80
C	Chair	200–300	1/10/81

Fig. 7.25

7.3.5 Keys

Composite keys

Simple keys and compound keys made up of more than one simple key have already been identified. Composite keys are keys made up of two or more attributes where one or more attributes of the key are not simple keys in their own right. Very often a second attribute has to be included to create a unique key. For example the key, book number/ chapter number, is a composite key since the attribute chapter number has no meaning without its associated book number.

Figure 7.25 provides data on equipment in a school, a school inventory. In order to identify uniquely an individual chair or desk, a composite key equipment code/identification number is required, written:

{equipment code/identification number}

In this case, identification number is not unique. Note that composite keys are shown enclosed by curly brackets.

Generated keys

Situations arise where it is necessary to generate a composite key. Consider the data structure, order, in Figure 7.26. Orders are not unique. Customer A47 has ordered ten of product number 210 on three

Customer number	Product number	Qty. ordered
A47	210	10
J21	479	15
A47	479	70
A47	210	10
A47	210	10
W15	298	5

Fig. 7.26

Customer number Sequence number	Product number	Qty. ordered
A47/1	210	10
J21/1	479	15
A47/2	479	70
A47/3	210	10
A47/4	210	10
W15/1	298	5

Fig. 7.27

occasions. A further attribute is required to provide a unique key; this could be date of order, time of order or more simply a sequence number for orders received from a given customer. The composite key, customer number/sequence number, provides a unique but generated composite key shown in Figure 7.27.

Implied keys

Figure 7.28 shows a data structure for staff with their project allocations. There is only an indirect dependency between project charge rate, project code and staff number. The project charge rate actually depends on the staff grade and project type. The 3NF data structure should be replaced by Figure 7.29, where the project code and staff number are replaced by their implied keys. In this case it is assumed a staff data structure links staff numbers to staff grades, and a project data structure links project codes to project types.

Normalisation	
UNF	3NF
Staff number Name Project code Project charge code	Staff number Name
	Staff number Project code Project charge rate

Fig. 7.28

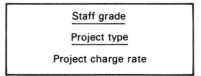

Staff grade
Project type
Project charge rate

Fig. 7.29

All staff of a given grade on a specific project type attract the same charge rate.

Aliases

Very often different descriptions exist for an identical attribute. Aliases should be documented as they are discovered to assist the task of merging data structures before creating the entity model in task 3.3.

Quotation number	Internal order number
Customer name Project number Quoted price	Customer code Customer name Product number Quantity ordered Delivery date

Fig. 7.30

The keys for the two data structures in Figure 7.30 were identified during the normalisation task. However, during systems analysis an alias was documented whereby internal order numbers correspond on a one-to-one basis with quotation numbers. The keys are identical and the data structures may be merged.

Key only data structures

The task of normalisation may occasionally yield a key only data structure in third normal form. The data structure in Figure 7.31 represents all the projects worked on by staff members.

The tests represented in Figures 7.21 and 7.22 should be carried out to determine dependencies between keys.

'For each staff number is there only one project number?' No.
'For each project number is there only one staff number?' No.

Normalisation	
UNF	1NF
Staff number Name Grade	Staff number Name Grade
Project number Project name	Staff number Project number ◄─┐ Project name ──┘
2NF & 3NF	
Staff number Name Grade	
Staff number Project number	
Project number Project name	

Fig. 7.31

Neither of the attributes can become the key and the key only data structure must be retained.

Task 3.2 may therefore create a large number of data structures in third normal form.

7.4 TASK 3.3: CREATE THE ENTITY DESCRIPTIONS

Task 3.3 takes all the data structures in third normal form and merges all those with identical keys. This creates a set of entity descriptions from which the entity model is created. Only data structures with identical keys may be merged. Merging is a very simple task accomplished by first

Customer number	Product number
Name	Depot number
Address	Qty. allocated
Product number	Product number
Depot number	Depot number
Qty. in stock	Location reference
Re-order qty.	Storage type
Customer number	Customer number
(delivery address)	Credit code
	Area code
	Representative

Fig. 7.32

Customer number	Product number
Name	Depot number
Address	Location reference
(delivery address)	Storage type
Credit code	Qty. in stock
Area code	Re-order qty.
Representative	Qty. allocated

Fig. 7.33

establishing all the different keys and secondly listing all the attributes for each key.

Figure 7.32 shows a number of third normal form data structures. There are only two different keys, the simple key, customer number, and the compound key, produce number/depot number. All attributes are listed under their appropriate key in Figure 7.33. The result is the identification of two entities along with their entity descriptions.

Occasionally merging of data structures may not be appropriate. For example consider Figure 7.34, a merged data structure obtained from two original data structures (Figure 7.35). In this case, only five per cent of quotations result in orders. A considerable amount of space could be saved by not merging the data structures. This is because there would be zero occurrences of the order data structure in ninety five per cent of cases.

The process of merging can introduce new data dependencies. The

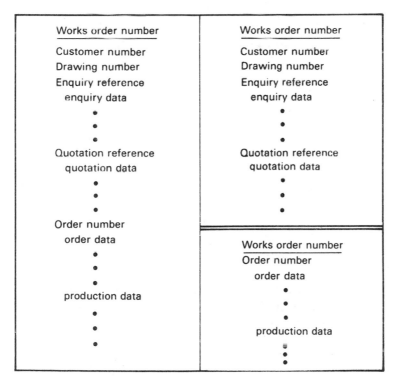

Fig. 7.34   Fig. 7.35

task of validation should be reapplied to the newly formed data structures. Task 3.3, therefore, identifies and provides for each entity an entity description.

7.5 TASK 3.4: CREATE THE ENTITY MODEL

The task of constructing an entity model from the entity descriptions is very straightforward and comprises five simple steps.

Step 1 introduces the entities on to the entity model by drawing a rectangle for each entity description. The rectangle is annotated with the entity name, and the entity key is written inside it (Figure 7.36 (a)). The entity is named order header and the key is a composite key customer number/order sequence.

Some of the entities may have entire keys which are composite keys. Each of the stand alone simple keys or inner composite keys within the composite key should be marked as a foreign key (Figure 7.36 (b)). This is step 2 and will allow relationships to be established at step 4. This step only applies to entities whose entire key is a composite key.

Step 3 requires that all compound keys are examined to ensure that an entity exists for each simple or composite key within the compound key. If any key does *not* exist, then an entity should be created with the appropriate key. The newly created entity should be marked as an operational owner (Figure 7.36 (c)).

Step 4 introduces relationships between entities. All entities with compound keys become members and must be connected to their owners. An owner relationship must exist for each element of the compound key, although it is acceptable to connect compound keys to compound keys at a lower level. The relationship may be to the owners via an entity with a compound key itself (see Figure 7.37). In this case, relationship (a) provides the owner keys customer number and product number; relationships (b) and (c) are, therefore, not required.

The final step is to introduce relationships connecting all foreign keys to their owners. Foreign keys may be found within the entity description or as a result of step 2 (Figure 7.36 (d)). As each relationship is introduced, its degree must be determined and entered on to the entity model.

By applying these simple rules, an entity model is built, bottom up, from the third normal form merged data structures.

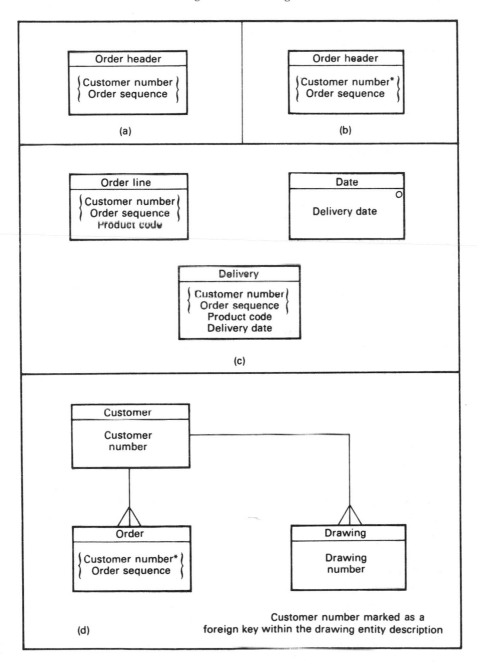

Fig. 7.36

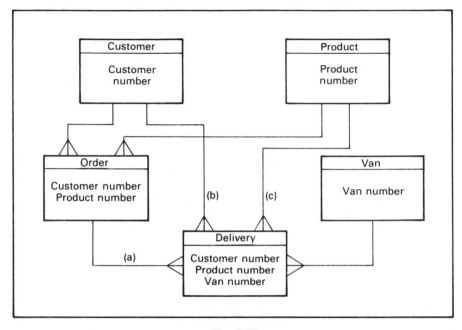

Fig. 7.37

7.6 TASK 3.5: CREATE THE LOGICAL ENTITY MODEL

The construction of the logical entity model from the stage A and stage B entity models again follows a number of steps – four in total. Each step is straightforward.

The initial step creates the first entities on the logical entity model by entering on to it all the operational owners from the stage A entity model. These operational owners were created to satisfy a specific user requirement and it is likely that they will be required as access routes into the database.

Operational owners from the stage B entity model result from missing elements of compound keys and key only data structures. These operational owners should only be entered on to the logical entity model if there are sound reasons for doing so, emanating from the user requirements.

Comparison of all remaining entities and insertion of further entities on to the logical entity model comprises step 2. It may be necessary to compare the entity descriptions at this step to identify logical aliases in

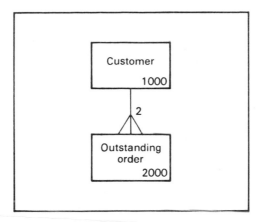

Fig. 7.38

entity names. Differences should be resolved by reference to the user requirements and all the entities inserted on to the logical entity model.

Step 3 now refers to relationships and commences with a comparison of all of them. Differences are again resolved by reference to the user requirements and the relationships are inserted on to the logical entity model.

The final step is to enter volumetric data on to the logical entity model. The number of logical occurrences of each entity should be determined and the entity box annotated as in Figure 7.38.

It may be necessary to estimate a minimum, maximum and expected growth rate. The numbers of occurrences may be added to the entity descriptions so that the entity description forms may be completed in task 3.6. Secondly, for each relationship, the average number of dependencies should be added to the logical entity model. Figure 7.38 shows a relationship between customer and order; the figure 2 should be interpreted as: 1000 customer entities have, on average, two orders outstanding, giving 2000 current outstanding order entities.

7.7 TASK 3.6: CREATE THE LOGICAL ENTITY DESCRIPTIONS

The final task in part 3 produces the logical entity descriptions by simple comparison of the entity descriptions from stages A and B. Differences are usually minor and resolution of them easy. However, at this task it is

Customer number	Name	Product number	Name	Qty.	Price
J2741	James	A273	Screws	100	1.00
		B147	Bolts	50	2.00
W1798	Waters	B147	Bolts	10	2.00
F1001	Frost	L110	Nuts	100	0.50
		A273	Screws	10	1.00

Fig. 7.39

possible to enhance the entity descriptions to include the entity size and the average number of occurrences. Thus the total amount of storage required for each entity can be calculated by simple arithmetic.

7.8 PART 3 SUMMARY

The base technique employed during part 3, logical data design, is the process of normalisation. Normalisation of data structures is necessary to ensure data structures are presented to the physical design stage in their simplest form, and to remove the possibility of loss of data integrity.

Re-examine Figure 7.10, a data structure for orders which is in unnormalised form. Figure 7.39 adds some data to the structure.

If a new customer negotiates an account, the unnormalised structure cannot record it until the customer places an order. Similarly, when the order for Frost is met, the information regarding Frost as a customer will be lost along with the product data regarding nuts. These are insertion and deletion problems with unnormalised data structures. Update problems also exist. For example, if the price of bolts is modified, then multiple update of the data structure is necessary since the price of bolts is recorded with each order for them.

The representation of data in third normal form overcomes insertion, deletion and update problems of this kind. Data structures are held in their simplest form and all duplication of data removed. The 3NF representation of the data structure in Figure 7.39 is shown in Figures 7.40 and 7.41. Note that the problems described are eliminated.

The six tasks of logical data design, part 3 are:
3.1 Selection of data structures for normalisation.

Normalisation	
UNF	1NF
Customer number Customer name	Customer number Customer name
Product number Product name Qty. Price	Customer number Product number ◄┐ Product name ──┤ Qty. Price ─────┘
2NF	3NF
Customer number Customer name	Identical to 2NF
Customer number Product number Qty.	
Product number Product name Price	

Fig. 7.40

3.2 Normalisation of each of the selected data structures using:
- Conversion of unnormalised data structures to first normal form including choice of appropriate keys
- Conversion of data structures in first normal form to second normal form
- Conversion of data structures in second normal form to third normal form
- Validation of the third normal form data structures.

3.3 Merging of third normal form data structures with identical keys to create entities and the entity descriptions.

3.4 A simple set of sub-tasks then transforms the entity descriptions into an entity model.

3.5 The entity model created during stage A is compared with the entity

Customer number	Name
J2741	James
W1798	Waters
F1001	Frost

Product number	Name	Price
A273	Screws	1.00
B147	Bolts	2.00
L110	Nuts	0.50

Customer number	Product number	Qty.
J2741	A273	100
J2741	B147	50
W1798	B147	10
F1001	L110	100
F1001	A273	10

Fig. 7.41

model from task 3.4; differences are resolved and the logical entity model is output to part 4.

3.6 The entity descriptions again created during stage A are compared with the entity descriptions from task 3.3. Anomalies are removed and a set of logical entity descriptions are output to part 4.

Chapter 8
Logical Process Design

8.1 INTRODUCTION

This chapter describes the fourth part of structured systems analysis and design, shown in Figure 8.1. Part 4 uses, as input, the entity function matrix, the entity life histories, the function descriptions, the on-line dialogue specification, the required physical DFDs, the logical entity model and the logical entity descriptions. These inputs are transformed into logical process outlines during part 4.

Part 4 commences with a review of the stage A documentation. The documents need to be updated to reflect the results of logical data design. The second task in part 4 enters all the required functions from the required physical DFDs into the process catalogue. All functions which have common processing requirements are collected together to form a process. Each process is then expanded into a detailed process outline – the third task – by reference to the entity function matrix, the entity life histories, the function descriptions, the logical entity model and the logical entity descriptions.

The logical process outlines, the logical entity model, the logical entity descriptions, the data dictionary and the on-line dialogue specification form input to part 5, physical design. The conclusion of part 4 marks the boundary between logical design and physical design. Part 5 must consider the target hardware, software and personnel.

8.2 TASK 4.1: REVIEW THE STAGE A DOCUMENTATION

The entity function matrix, the entity life histories and the function descriptions from stage A may require updating to reflect changes to the entity descriptions and entity model, as a result of part 3.

Part 3, logical data design, very often produces new entities as well as modification or even deletion of existing entities. Task 4.1 is to bring the entity function matrix up to date with the logical entity model. The

241

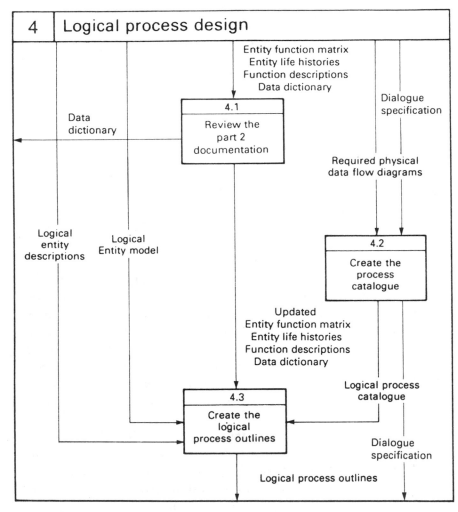

Fig. 8.1

entity function matrix must have a new row added for each new entity with the effects of functions on that entity charted. In addition, an entity life history must be constructed for each new entity.

The addition of new entities on to the entity function matrix is a relatively easy process. More problems are encountered if entities are merged or divided as part of logical data design.

Figure 8.2 shows a section of an entity model from stage A. Figure 8.3

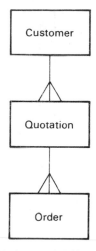

Fig. 8.2 Entity model (from stage A)

shows the same section of the logical entity model from part 3. Note that there is one new entity and a merging of two entities. The new entity, product, has emerged during part 3 and has been included on the logical entity model. In this case, in contrast to earlier, the entities quotation and order on the stage A entity model have been merged into one entity, order, on the logical entity model. This results from merging data structures with identical keys during part 3.

The entity function matrix, Figure 8.4, shows the three entities from Figure 8.2.

The entity descriptions for quotation and order are shown in Figure 8.5. In third normal form, four data structures result, as shown in Figure 8.6. Quotation numbers are unique within the system.

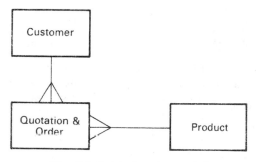

Fig. 8.3 Logical entity model

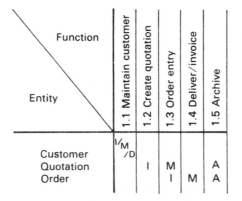

Fig. 8.4

Quotation

UNF, 1NF, 2NF	3NF
Quotation no.	Quotation no.
Customer no. ◄─┐	Customer no.*
Customer name ─┘	Product code*
Product code ◄─┐	Quoted price
Product description ─┘	
Quoted price	

Customer no.

Customer name

Product code

Product description

Order

UNF, 1NF, 2NF	3NF
{ Customer no.	Quotation no.
{ Order no.	Customer no.*
Quotation no. ◄─┘	Order no.
Qty. ordered	Qty. ordered

Fig. 8.5

Quotation:	Quotation no. Customer number* Product code* Quoted price	Customer:	Customer number Customer name
Product:	Product code Product description	Order:	Quotation number Customer number* Order number Quantity ordered

Fig. 8.6

The order entity key, after the first three steps of the normalisation task, is a compound key comprising a composite key, customer number/order number, and a simple key, quotation number. The final validation of the data structure revealed that the composite key customer number/order number was dependent on the quotation number. The simple key, quotation number became the key to both the quotation entity and the order entity. The entities quotation and order were, therefore, merged in the construction of the logical entity model shown in Figure 8.3.

The updated entity function matrix must merge the two existing rows, quotation and order from Figure 8.3. The new matrix must also have a new row for the product entity together with a new function column, function 1.6, to maintain the product entity. Figure 8.7 shows the updated entity function matrix. Note also that the product entity is read by functions 1.2, 1.3 and 1.4.

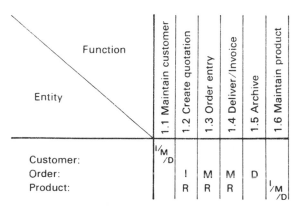

Fig. 8.7

The entity life histories for the order and quotation entities need to be redrawn as one entity life history. In this case it is likely that the two histories will follow in sequence since orders always follow quotations, that is if an order is received. A new entity life history is required for the product entity.

Finally, all status indicators should be reviewed. These form a very important integrity check during processing. They are carefully recorded on the logical process outlines during task 5.3.

8.3 TASK 4.2: CREATE THE PROCESS CATALOGUE

The required physical DFDs indicate a mode of processing for each function. The modes were selected by the user during part 2. The mode might be simply on-line or batch. The mode may be further divided into on-line update, on-line enquiry, batch report or batch update and further by time period and data access requirements.

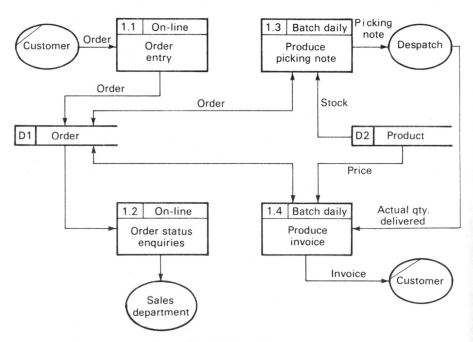

Fig. 8.8 Level 2

Mode/process no.	Function reference	Name
On-line enquiry 1	1.2	Order status enquiry
On-line update 2	1.1	Order entry
Batch update-daily 3	1.3 1.4	Produce picking note Produce invoice

Fig. 8.9 Logical process catalogue

The mode may be batch update weekly or batch update daily or batch update weekly stock file or batch update weekly sales ledger. The mode represents the exact resources required by the process.

A catalogue of processing modes can be established with each function being posted to one mode in the catalogue. Very often the level 2 DFDs represent functions which do not comprise more than one mode of processing. It is these functions which can be posted to the catalogue. The process catalogue shown in Figure 8.9 resulted from the posting of functions represented on the level 2 required physical DFD shown in Figure 8.8.

All the functions that have the responsibility box, indicating computer processing, are entered into the logical process catalogue. The catalogue provides a list of all required computer processing. As each function is entered into the process catalogue, the designer should check that the function is fully understood and documented since it is that documentation which will form the basis of the process outlines.

The objective of posting functions into the process catalogue is to identify groupings of functions which may become program modules. If the processing modes chosen are simply on-line and batch, then perhaps two programs would result, one on-line and one batch. The decision tree below shows a more practical set of processing modes.

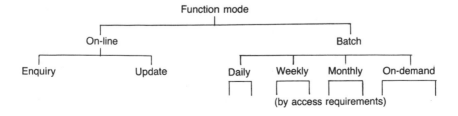

Therefore, the catalogue modes may be:
- on-line enquiry
- on-line update
- batch daily sales area
- batch daily product area
- batch weekly
- batch monthly sales area
- batch monthly product area

where sales and product areas are logical sections of the database.

8.4 TASK 4.3: CREATE THE LOGICAL PROCESS OUTLINES

The input to task 4.3, creation of the logical process outlines comprises the updated entity function matrix, entity life histories, logical function descriptions together with the logical entity model, logical entity descriptions, and the logical process catalogue. The logical process outlines will be used during part 5 to create program specifications.

Only processes listed in the process catalogue require logical process outlines. All processes that are to remain manual will have been excluded from the catalogue. The creation of logical process outlines adds the fine processing detail, including the processing of entities to the set of function descriptions for the process. A process outline is both detailed and logical. A logical process outline should be created for each on-line function. The logical process outlines will be linked to process complete transactions during part 5.

A logical process outline should be created for each batch process. A process may combine a number of functions, all of which share the same timescales and the same resource requirements. These two statements constitute the rules for the relationship between logical process outlines and functions. These rules follow on from the posting of functions into the process catalogue. There are, however, other considerations which may influence the number of logical process outlines produced.

Many current target systems provide software to link programs or modules in order to process a complete transaction. The implementation of the dialogue from screen formats through processing is, therefore, dependent on the target system. In this instance a logical process outline should be produced for each on-line function. The functions should then be linked to form transaction processing routines.

The only potential problem with batch logical process outlines is the size. If very many functions are catalogued as batch daily, requiring access to a particular section of the database, then some further splitting may be necessary and appropriate.

The logical process catalogue shown in Figure 8.9 would result in one batch update daily process outline. This is reasonable since both functions are batch daily and require access to the orders and products data stores. It could be argued, however, that since the two functions access different orders, and since function 1.3 updates the product data store and function 1.4 only references the product data store, then two logical process outlines should result.

As the methodology progresses nearer and nearer to the production of physical specifications, which depend on a knowledge of the target hardware and software, so the methodology becomes much more guidelines rather than rules. The grouping of functions into processes is an example of a guideline rather than an exact rule.

Each process outline comprises a process outline heading and a number of operation entries. The heading provides documentation of the complete process including the process number, a process name, the process mode, its frequency, the volume of transactions plus a brief description of the processing.

Operation entries may take many forms. Operations may perform the modification of an entity, validate some input, format some output or perform a calculation. For each operation, the documentation should include some or all of the following: operation number, the name of the entity affected by the operation, the 'valid previous' and 'set to' status indicators for the entity, any input or output or error handling, plus a detailed description of the processing associated with the operation.

Figure 8.10 shows a blank process outline form.

Each process will affect a number of entities. The entities affected by a process and the effect on the entity can be obtained by reading down a column on the entity function matrix. A column of an entity function matix is shown in Figure 8.11.

The function, produce invoice, reads contract, modifies order line, modifies stock and inserts an invoice and an invoice line, although the sequence of effects on the entities cannot be determined. The sequence of effects is made clear by reference to the function description.

Further, by reference to the entity life histories, the entity status that must exist for the functions to execute validly may be determined.

LOGICAL PROCESS OUTLINE

SYSTEM:	DATE:
AUTHOR:	PAGE: of

PROCESS No: NAME:

MODE: FREQUENCY: VOLUME:

BRIEF DESCRIPTION:

DFD Functions.

ENTITY NAME						
EFFECT						
VALID PREV.						
SET TO						

Op. no.	Entity Name	Effect	Status Ind. Valid prev.	Set to	Description Narrative	Ref.	I/O ref.	Error ref.

Fig. 8.10

	2.5 Produce invoice
Contract	R
Order line	M
Stock	M
Invoice	I
Invoice line	I

Fig. 8.11

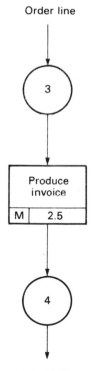

Order line

Produce invoice

| M | 2.5 |

Fig. 8.12

Secondly, again from the entity life histories, the 'set to' status on completion of successful processing may be determined. An extract of the entity life history for the entity order line is shown in Figure 8.12.

The order line entity must have status 3 for the function, produce invoice, to be valid. Further, after the function has completed successfully, the status should be set to 4. This will allow a subsequent function to archive the order line. The effect on each entity of the function can be described by one of the operations, Read, Insert, Modify, Delete or Archive, and the status indicators 'valid previous' and 'set to'.

The entities affected by the process should be identified from the entity function matrix, perhaps by reading down several function columns. For each entity affected, the entity life history should be examined to determine the 'valid previous' and 'set to' status for each of the functions within the process. This information is extracted as a first step towards the completion of a logical process outline.

The extract is a number of columns from the entity function matrix, enhanced with status information from the appropriate entity life histories. It provides a concise reference of the process' effect on the

LOGICAL PROCESS OUTLINE

SYSTEM: *SALES ACCOUNTING*	DATE:
AUTHOR: *G. CUTTS*	PAGE: *1* of *2*

PROCESS No: *1*   NAME: *PRODUCE INVOICE*

MODE: *BATCH*   FREQUENCY: *WEEKLY* VOLUME:

BRIEF DESCRIPTION: *Insert invoice & invoice lines in to the database for all delivery notes created during the previous week. Note the invoice is not printed by the process.*

ENTITY NAME	Contract	Order line	Stock	Invoice	Invoice line	
EFFECT	R	M	M	I	I	
VALID PREV.	—	3	1	—	—	
SET TO	—	4	1	1	1	

Fig. 8.13

entities, a vital section of the process outline. This concise reference is included as part of the logical process outline itself (see Figure 8.13).

The detail of each operation's processing may be expressed in English, structured English, decision tables or trees, pseudo-code, or some appropriate combination of techniques. In describing processing in detail, error conditions must be considered. Here again, the updated entity life histories are important as they document the effect of abnormal events on each entity.

Returning to the process produce invoice, the function description from stage A was: 'For all advice notes produce an invoice line. If it is the first invoice line enter an invoice header. Prices are obtained either from the contract entity or the stock entity. Update the stock entity to record the quantity despatched.'

The initial logical process outline for produce invoice is shown in Figure 8.14. This is constructed by using the function descriptions to sequence the entries in the reference section of the process outline. The processing to handle the effect on the entities must now be described, together with any intermediate processing such as calculation of the invoice value, the handling of error conditions and the validation of input documents.

The specification of error processing is often more complicated than the specification of standard processing. The use of error references in logical process outlines removes the necessity to include detailed error processing within the body of the standard process description. All error processing can be collated and specified separately in an error processing logical process outline. An example of the use of an error reference is shown in Figure 8.15.

The description refers to error reference E100 where a description of the error processing can be found. This process outline will eventually be programmed with a call to the error processing module, reference E100.

To complete the process outline, reference needs to be made to any system input or output. An operation such as print sales statement will involve access to several entities as well as producing the printed sales statement. A complete specification of the input or output format within the process is not necessary; a reference to the format is required. A column input/output reference is included on the process outline document. Figure 8.15 shows an input reference. Input reference I100 contains the physical format of an advice note.

LOGICAL PROCESS OUTLINE

SYSTEM: *SALES ACCOUNTING*	DATE:
AUTHOR: *G. CUTTS*	PAGE: *1* of *2*

PROCESS No: *1*	NAME: *PRODUCE INVOICE*
MODE: *BATCH*	FREQUENCY:*WEEKLY* VOLUME:

BRIEF DESCRIPTION: *Insert invoice & invoice lines into the database for all delivery notes created during the previous week. Note the invoice is not printed by the process.*

ENTITY NAME	Contract	Order line	Stock	Invoice	Invoice line	
EFFECT	R	M	M	I	I	
VALID PREV.	—	3	1	—	—	
SET TO	—	4	1	1	1	

Entity		Status Ind.	
Name	Effect	Valid prev.	Set to
Order line	M	3	4
Stock	M	1	1
Contract	R		

Fig. 8.14(a)

LOGICAL PROCESS OUTLINE

SYSTEM: *SALES ACCOUNTING*	DATE:
AUTHOR: *G. CUTTS*	PAGE: *2* of *2*

PROCESS No: *1*	NAME: *PRODUCE INVOICE*
MODE: *BATCH*	FREQUENCY: *WEEKLY* VOLUME:

BRIEF DESCRIPTION: *Insert invoice & invoice limes in to the database for all delivery notes created during the previous week. Note the invoice is not printed by the process.*

ENTITY NAME	Contract	Order line	Stock	Invoice	Invoice line	
EFFECT	R	M	M	I	I	
VALID PREV.	–	3	1	–	–	
SET TO	–	4	1	1	1	

	Entity Name	Effect	Status Ind. Valid prev.	Set to
	Invoice	I	–	1
	Invoice line	I	–	1

Fig. 8.14(b)

LOGICAL PROCESS OUTLINE

SYSTEM: *SALES ACCOUNTING*	DATE:
AUTHOR: *G. CUTTS*	PAGE: *1* of *2*

PROCESS No: *1* NAME: *PRODUCE INVOICE*

MODE: *BATCH* FREQUENCY: *WEEKLY* VOLUME:

BRIEF DESCRIPTION: *Insert invoice & invoice lines in to the database for all delivery notes created during the previous week. Note the invoice is not printed by the process.*

ENTITY NAME	Contract	Order line	Stock	Invoice	Invoice line	
EFFECT	R	M	M	I	I	
VALID PREV.	–	3	1	–	–	
SET TO	–	4	1	1	1	

Op. no.	Entity Name	Effect	Status Ind Valid prev.	Set to	Description Narrative	Ref.	I/O ref.	Error ref.
1					Read the advice note details & validate the input fields.	V10	I100	
					If validation error reject.			E100
2	Order line	M	3	4	Obtain the relevant Order line. If <u>not found</u> reject the advice note			E200
					If <u>found but status indicator errors</u> then reject the advice note.			E300
					Modify the order line status to indicate Order Line despatched.			
3	Stock	M	1	1	Obtain the relevant stock item.			
					If <u>not found</u> report error			E210
					If <u>found but status indicator errors</u> then error.			E310
					Extract the stock item price.			
					Modify the entity to de-allocate the stock & also to decrement the qty. on hand.			
4	Contract	R			Read the relevant contract for all order lines stating a contract number.			
					If not found error.			E220
					continued			

Fig. 8.15(a)

LOGICAL PROCESS OUTLINE

SYSTEM: *SALES ACCOUNTING*	DATE:
AUTHOR: *G. CUTTS*	PAGE: *2* of *2*

PROCESS No: *1*   NAME: *PRODUCE INVOICE*

MODE: *BATCH*   FREQUENCY: *WEEKLY* VOLUME:

BRIEF DESCRIPTION: *Insert invoice & invoice llimes in to the database for all delivery notes created during the previous week. Note the invoice is not printed by the process.*

ENTITY NAME	Contract	Order line	Stock	Invoice	Invoice line
EFFECT	R	M	M	I	I
VALID PREV.	—	3	1	—	—
SET TO	—	4	1	1	I

Op. no.	Entity Name	Effect	Status Ind. Valid prev.	Set to	Description Narrative	Ref.	I/O ref.	Error ref.
5					Extract the contract price. If not contract use the price from the stock item entity. Calculate the invoice line value = quantity delivered (delivery note) x price.			
6	Invoice	I	—	1	If this is the first advice note for a given customer insert an invoice entity into the database.			
	Invoice line	I	—	1	Insert an invoice line entity into the database.			

Fig. 8.15(b)

Note that the error references have a leading significant digit. The one hundred series represents validation errors, the two hundred series represents database errors and the three hundred series status indicator errors. Strictly speaking, the error references E210 and E310 refer to entity not found in this specific case. It is good practice to check the status indicator before any operation on an entity, with the exception of read. Reference V10 provides the detailed validation rules for the advice note.

Where the narrative description is insufficient to provide the exact detail, reference may be made to a more detailed description. This may be in the form of a decisions table or tree, it may be an algorithm specified in pseudo-code or it may be a set of validation rules. This reference is shown along with the narrative description of the operation. Figure 8.15 includes a reference V10 to the validation rules for an advice note input in format I100.

A logical process outline may make reference to other logical process outlines for common routines and error processing, for input and output formats, and for documents supporting the narrative description of an operation for validation or detailed calculation rules.

A set of logical process outlines with supporting documentation is the major deliverable.

8.4.1 Specifying process logic

The examples in this chapter use a structured narrative to describe each operation of the logical process outline. When this outline is transformed into a program specification, coded and tested, it is the task of the program specifier and programmer to take into account the target hardware and software.

The tendency for specifiers of operations on logical process outlines is to use a familiar format often based on the specifier's experience. The author's experience is very much Cobol in a third generation environment using a Codasyl database. The examples demonstrate this background.

Many programmers use a fourth generation language exclusively. The style of operation specification may be varied to suit the programming environment. For example, the operations in Figure 8.15 may be specified as follows:

```
GET         advice note details and validate the input fields
IF          validation error reject
SELECT      order line
LOCATE      relevant order line
IF EOF      reject the advice note
REPLACE     order-line-status with 4
etc.
```

or instead of LOCATE

```
SELECT   order-line
FROM     order-line-entity
WHERE    order-line-key = delivery-note-key
```

In the cases above, the use of the fourth generation language verbs within the operation specification is quite acceptable.

Finally, the emergence of code generators leads to another style of operation specification. In this case the language of the code generator might be used as the specification language to ease the program generator's translation task.

The specification of process logic will very often make reference to entities and attributes of entities. Care should be taken to ensure all entity names are identical to the names on the logical entity model. Similar care should be taken to ensure all attribute names are identical to the names used in the logical entity descriptions and to those entered into the data dictionary.

One convention for attribute naming is to use the dot notation, e.g. customer.credit-limit where credit-limit is an attribute of the entity customer.

8.5 PART 4 SUMMARY

Part 4 concludes logical design before physical design takes place in part 5. There are three tasks within part 4. Task 4.1 reviews the entity life histories, the entity function matrix and the function descriptions, produced during stage A, against the logical entity model produced during part 3. The DFDs represent the required processing and the mode of processing. The functions are posted to a logical process catalogue, task 4.2. This documents, for each function, its processing mode, batch or on-line. All functions remaining manual are excluded from the catalogue. The catalogue is then used by task 4.3 to provide a list of processes for which logical process outlines should be created.

The logical function descriptions specify, in business terminology, the detailed processing for each function. These are now transformed into logical process outlines by task 4.3. The effect on each entity is documented by reference to the entity life histories and reference is made to validation and error processing rules.

Finally, the logical process outlines are validated against the user's view of the system described by the business specification in the stage B user review.

Chapter 9
Physical Design

9.1 INTRODUCTION

This chapter describes stage C and part 5, physical design, shown in Figure 9.1. Stage C uses the logical process outlines, the logical entity model, the logical entity descriptions, the data dictionary and the on-line dialogue specification to create a complete, detailed, physical design specification. The design specification comprises a file or database specification, a set of program specifications, an implementation plan, an operations manual and a user manual.

The input to stage C is the logical design specification. This is independent of the target hardware and software. Stage C transforms the logical specification into the physical specification by taking into account the constraints imposed by the target system. The output from the stage is, therefore, installation dependent.

Moreoever, the design must adhere to objective measures of performance and resource usage. The tasks within the stage must provide for performance and resource usage estimation, with the opportunity for design tuning if the objectives are not met.

9.2 TASK 5.1: SPECIFY THE PHYSICAL FILES OR DATABASE

The objective of this task is to create the physical specification of the files or database and the physical specification of the records and record content. The data dictionary should now be fully completed as part of the specification.

The initial step is to annotate the logical entity model with a series of arrows to show the entry points into the model. These arrows will lead to index tables or indexed files allowing direct access to the data. Operational owners indicate entry points identified during the earlier stages.

Figure 9.2 shows a logical entity model with entry points marked by

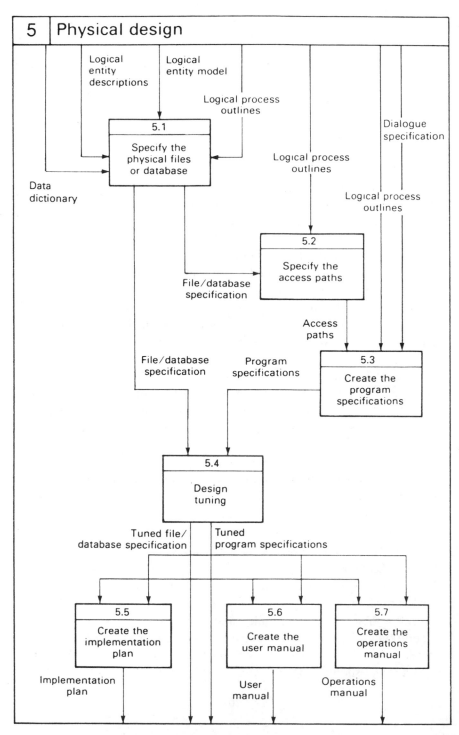

Fig. 9.1

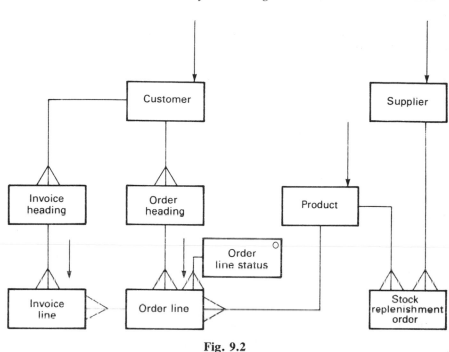

Fig. 9.2

arrows. Note that all entities at the head of a hierarchy and all entities with operational owners must be entry points. The entry points marked on the customer, product and supplier entities result from the entities being at the head of a hierarchy. The entry point on order line results from the operational owner, order line status.

The entry point on the invoice line entity results from a requirement to satisfy a process outline. The process outline stated: 'Read all the invoice lines inserted into the system over the past week, and produce a report of the total invoice value delivered each day.' Note that an invoice line represents a delivery against an order line.

Entry points may result, therefore, from the requirements of a process outline. All the process outlines must be validated against the logical entity model to determine entry points required. Many of the entry points required by the process outlines will have been indicated already, since the entities will be either at the head of a hierarchy or be linked to an operational owner.

Subsequent steps in task 5.1 depend on the target system. A set of guidelines for creating the physical specification should be constructed for each target system. Some examples are given below. These examples are given as an indication of the steps involved when conversion is required to indexed sequential files, or to network or relational schema. The examples are not intended to be an exact set of rules.

9.2.1 Conversion to indexed sequential files

The guidelines for creating a set of indexed sequential files follow a similar route to those used to create logical files in stage A. The guidelines are explained by reference to the logical entity model shown in Figure 9.2.

Guidelines

Commence a new file for each entity at the head of the hierarchy. Three initial files result from Figure 9.2: a customer file, a product file and a supplier file. Follow the relationships from the entities already placed in a file to determine the member entities.

The member entity may be treated in one of three ways:
(1) The entity may become a repeating group within the owner member entity record
(2) The entity may become a separate record type in the owner member entity file
(3) The entity may create a new file.
Some entities, for example the stock replenishment order, have a choice of owner entity file for ways (1) and (2) above. If way (1) is agreed, then either a fixed maximum number of repeats must be decided or variable record lengths must be used.

Figure 9.2 has many member entities related to owner entities, customer, product and supplier. Reference to the process outlines indicates that the stock replenishment order entity is processed on-line with product entities. For this reason stock replenishment orders are included as a separate record type in the product file.

The customer entity possesses two members, order heading and invoice heading. Since there are many repeat orders and, therefore, many invoices in this system, way (3) is chosen for both cases. The invoice heading entity and order heading entity will create new files, effectively making them into hierarchy heads.

The ratio of invoice headings to invoice lines is high, whereas the ratio

File	Record types	Key
Customer	Customer	Customer code
Supplier	Supplier	Supplier code
Product	Product	Product number
	Stock replenishment order	Product number/supplier code
Order	Order heading	Customer code/order reference
Order line	Order line	Customer code/order reference/product number*
Invoice	Invoice heading	Customer code/invoice number
	Invoice line	Customer code/invoice number/delivery note number**

* Order line status is included as an attribute of the entity Order line.

**Order reference and Product number are included as attributes of Invoice line.

Fig. 9.3

of order headings to order lines is very low, sometimes one to one. The overhead of processing invoice headings when accessing invoice lines only is therefore low, and for that reason the invoice line entities are included in the invoice heading file. Order lines are located in their own file.

The final file specification is shown in Figure 9.3. Each key is obtained from the logical entity description. Note that the access points shown on the logical entity model are preserved.

9.2.2 Conversion to a network schema

The conversion step is relatively straightforward. For each entity on the logical entity model, three parameters must be specified, since entities become records. The three parameters are:

Entity or record name   A unique name to be used in the schema
Record identifier   A unique number to identify the record type
Location mode   The storage method

Generally, there are two choices of location mode, CALC and VIA. With location mode CALC, the record key is randomised to derive a page number where the record will be stored. The record will occupy the first available line in the page and is linked to the CALC set for the

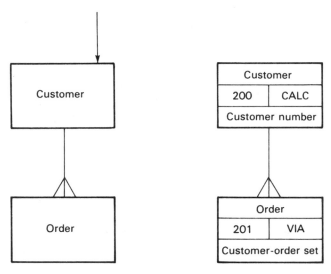

Fig. 9.4

page. This means that the record may be accessed directly. Entities with location mode VIA an owner record require the record to be stored as near as possible to the current occurrence of its owner. This allows member records to be clustered around their owners and access is generally, but not exclusively, via their owner. An owner/member relationship is called a set.

Figure 9.4 shows two entities and their record identifiers.

The CALC location mode is used for all entities marked as entry points into the database. Where an entity can be 'located via' more than one owner, the choice may be made by reference to processing requirements. This is similar to the choice made for the stock replenishment order in the previous example.

The record identifiers for the logical entity model, shown in Figure 9.2, are:

Logical entity	Record name	No.	Location mode	Access key or set
Customer	CUST	200	CALC	CUSTOMER-CODE
Product	PROD	201	CALC	PRODUCT-NUMBER
Supplier	SUPP	202	CALC	SUPPLIER-CODE
Stock replenishment order	SRO	203	VIA	PROD-SRO-SET
Order heading	OHEAD	204	VIA	CUST-OHEAD-SET
Order line*	ORD	205	CALC	ORD-LINE-STATUS
Invoice heading	IHEAD	206	VIA	CUST-IHEAD SET
Invoice line*	INV	207	CALC	DATE-OF-DELIV

* Because of the chosen access key directly related to the processing, duplicate entity occurrences are likely which are uniquely identified by the attributes or fields within the entity.

Six parameters must be specified for each relationship. The term set is used for a relationship.

Set name	A unique name that will be used in the schema.
Set order	This determines where a new member will be linked in the set and thus determines the sequence of member records.
Data name	This is only required for sorted members. It is the data name within the record for sequencing the members.
Pointers	A definition of the links required from owners to members and members to members.
Membership rules	These allow the specification of storage class and are necessary for optional links.
Duplicates	These are only required for sorted members and provide a definition of the processing required if duplicate keys exist.

There are seven relationships in Figure 9.2. The product stock replenishment order relationship might be specified as:

Set name	PROD-SRO
Set order	LAST
Data name	–
Pointers	NEXT
Membership rules	MANDATORY AUTOMATIC
Duplicates	NOT ALLOWED

This short example is not intended to provide any detail; it serves only to show that guidelines can be constructed for converting from the logical entity model to a schema.

In both cases entities have become records, either as record types within an indexed sequential file or as database records. The logical entity descriptions form the basis of the record descriptions.

The additional information required to convert a logical entity description into a record description is the physical data relating to the size and format of each attribute. This can be obtained from the data dictionary. Volumetric data is also required to enable the file size to be calculated from the record size.

Some careful arrangement of attributes is necessary if more than one record type is required in a file. Record type indicators are required along with a common key position and format. Each record must include a status field to hold the current status value from the entity life histories.

9.2.3 Conversion to a set of relational tables

The conversion is very straightforward. Since the logical entity model is in third normal form, each entity can be realised as a relational table. The relationships from the logical entity model indicate potential joins between the relational tables.

The conversion from logical to physical requires a detailed knowledge of the target hardware and software and, in this case, a detailed knowledge of the installation's database standards and practice.

9.3 TASK 5.2: SPECIFY THE ACCESS PATHS

The objective of this task is to specify the access path for every operation involving access to an entity, on every logical process outline. Each process outline is processed in turn, against the physical file or database specification.

The logical process outline will state the entity name, the effect, the 'valid previous' status and the 'set to' status. The method of access must state exactly how the effect is achieved on the named entity.

Figure 9.5 shows an extract from a logical process outline; only those operations which affect an entity are shown.

Operation no.	Entity name	Effect	Description
			Repeat the following processing for all order lines.
1	Order line	R	Check the Order-lines status. If the status = delivered then next-order line (operation 1) else (operations 2-7)
2	Product	R	If qty. ordered > qty. on hand then (operation 3) else (operations 4-7)
3	Order line	M	Change status to pending
4	Product	M	Subtract qty. order from qty. on hand
5	Order line	M	Change status to delivered
6	Invoice heading	R	If there exists an Invoice heading then (operation 8) else (operations 7-8)
7	Invoice heading	I	Create an Invoice heading
8	Invoice line	I	Create an Invoice line

Fig. 9.5

Figure 9.6 shows the access paths using the indexed sequential files.

Figure 9.7 shows the access paths using a network database. If the database is being implemented as a set of relational tables, the access paths may be specified simply by using, for example, SQL.

Note that Figures 9.6 and 9.7 are dependent on the hardware and software of the target system.

9.4 TASK 5.3: SPECIFY THE PROGRAMS

The program specifications form part of the documentation handed over from the development team to the implementation team. A program specification should provide a clear, complete, correct and unambiguous description of a program to enable a programmer to design, code and test the program.

Operation no.	Entity name	Effect	Access path
1	Order line	R	Read (next)
2	Product	R	Set the access key to the product number from the order line Search & read
3	Order line	M	Rewrite (of previous read)
4	Product	M	Rewrite (of previous read)
5	Order line	M	Rewrite (of previous read)
6	Invoice heading	R	Set the access key to customer code from the order line and invoice number to this week's number Search & read
7	Invoice heading	I	Set access key to customer code from the order line and invoice number to this week's number Write (an invoice heading)
8	Invoice line	I	Set access key to customer code from the order line, invoice number to this week's number and delivery note number to that from the input Write (an invoice line)

Fig. 9.6

The format of program specifications is installation dependent. A minimum program specification should comprise:
• the program number and name
• an overview or short description
• the data access required
• input, output and screen formats
• menu and dialogue design and formats
• the processing required

9.4.1 On-line programs

The logical process outlines together with the physical access paths provide the substantial element of the physical program specification.

Operation no.	Entity name	Effect	Access path
1	Order line	R	Set the access key, order line status to delivered Obtain first (for 1st access only) Obtain next (for subsequent access)
2	Product	R	Set the access key, product number, to the product number from the order line Obtain calc.
3	Order line	M	Store (of previous obtain)
4	Product	M	Store (of previous obtain)
5	Order line	M	Store (of previous obtain)
6	Invoice heading	R	Set the access key for the Customer entity to the customer code from the order line Find (the customer) Obtain first Ihead in Cust-Ihead set while there exists invoice headings and invoice heading not found If invoice-number = this week's number then invoice heading found else obtain next Ihead in Cust-Ihead set
7	Invoice heading	I	Set invoice number to this week Store Invoice heading
8	Invoice line	I	Set delivery note number to input number Store Invoice line

Fig. 9.7

Logical process outlines reference the input specifications, the output specifications, the on-line dialogue specifications and the data dictionary. Screen designs and report designs are required to complete the program specifications.

Many target systems will possess software for screen and report design and documentation; some will possess software for menu and dialogue creation. The creation and execution of menu screens is becoming available on many systems. The final on-line program specification will therefore comprise:

- the logical process outline
- the access paths
- screen designs, validation rules and error messages, report layouts
- the menu and dialogue designs

as a set of individual documents or in a represented format conforming to the installation standards.

The only step then required to complete the on-line program specifications is the identification of common processing. Any common processing not identified during the production of the logical process outlines should be extracted from the program specifications and collected into a common module specification.

9.4.2 Batch programs

Many batch programs, such as the on-request production of standard reports, are free-standing. Very often there exists a necessary or preferred sequence of execution of batch programs. Sequencing may be necessary to extract information prior to sorting and printing. Similarly, it may be necessary to execute a file update program prior to using the file for the production of reports. The sequencing of programs into batch run flows can be determined by reference to the required physical DFDs.

The definition and sequencing of batch programs inevitably leads to the identification of temporary files and transaction files. For example, if the section of the DFD shown in Figure 9.8 is transformed into a batch

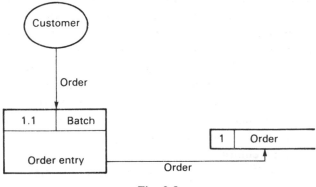

Fig. 9.8

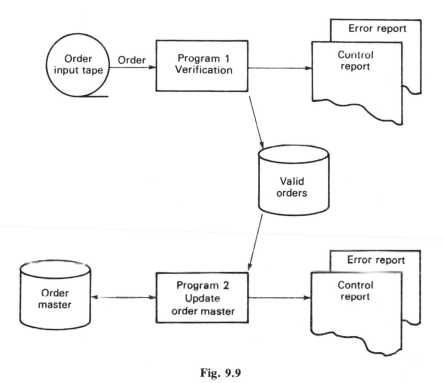

Fig. 9.9

program sequence, the program sequence shown in Figure 9.9 may result. Program 2 results from a logical process outline; program 1 requires its processing to be defined.

Program specifications, which enable batches of data to be processed, need not necessarily result from logical process outlines; they may result from the definition of batch run flows. In this case, program 1 simply verifies the accuracy of the data preparation; the order entry defined by function 1.1 on the DFD is translated into a logical process outline and then into a program specification.

The final batch program specifications will comprise:
- the logical process outline or a description of the processing required
- the access paths (possibly including temporary or transaction files)
- input and report layouts.

The batch run flow will be documented in task 5.7, creation of the operations manual.

9.5 TASK 5.4: DESIGN TUNING

Design tuning is necessary to ensure the performance and resource usage objectives are met.

Performance objectives include batch run times, transaction response times and recovery times. The associated issues of security and privacy, which both affect performance, are also included. Resource objectives include direct access device space and device utilisation.

The objectives were specified during stage A. These objectives now require to be developed into a set of detailed physical objectives against which the final system can be measured. Resource objectives are generally easier to specify than performance objectives. The amount of direct access space and the number of devices, e.g. terminals, is very often fixed for a particular system.

This task requires that the performance of the major on-line and batch programs be calculated in order to check if the design objectives have been met. If they have not been met, design modifications may be necessary to ensure that they are. If objectives are shown not to be achievable, then modifications to them may be necessary. In either case, the reworking of earlier stages and tasks will result.

Design tuning is not a self contained task; it is the calculation of the expected performance and resource utilisation for the current system design, and the comparison of the results against the design objectives.

The utilisation of resources is easier to calculate than system performance. The calculation of batch run times is particularly difficult. It is often necessary to involve the hardware and software suppliers to provide performance figures. In many situations, the only reliable figures result from actual measurement of the system performance.

Design tuning is not just a task which iterates through earlier development stages; it is a task to which the implementation team should return as the implementation progresses and more figures on actual performance and utilisation become available.

9.5.1 Ways of improving performance

If the performance of the system as calculated or as observed does not meet the objectives, then ways of improving it may have to be found. The simple introduction of more hardware may not solve the problem. The design itself may require modification.

The design of the files or database should be examined in detail. Changes to the indexing methods, file and record sizes, are all examples of detailed changed which may affect performance. Other changes include the duplication of data to allow concurrent access and the reduction of the amount of data held on-line to reduce the on-line storage requirements. Many changes do not affect the user's requirements, many do.

A reduction in the level of service offered to the user by the new system may be the only way to improve performance. The removal of direct access to entities using an alternative key, and the removal of direct access to be replaced by extract, sort and print and a subsequent change from on-line to batch mode, are examples of reduced service levels. These changes will improve the performance of the remaining on-line modules.

Task 5.4 is the final task relating to the data and program specifications before they are handed over to the implementation team. Design tuning provides an opportunity for the development team to review their design against the system objectives before detailed implementation work commences. Errors and potential problems identified at this stage are easier and less expensive to correct than errors detected during program testing, system testing, user testing or live operation.

9.6 TASK 5.5: CREATE THE IMPLEMENTATION PLAN

The development team should have gained valuable knowledge of the system and the user during the development stages. The team should prepare a detailed implementation plan based on their experience, to hand over to the implementation team.

Implementation includes the design, writing and testing of the programs, system testing, user testing, changeover, review and preparation for maintenance and modification. The method by which plans are constructed and presented is again installation dependent. Some installations use sophisticated planning and monitoring software, others use simple bar charts.

All methodologies should provide a starting point for the next phase of the project. The implementation plan should provide a planning start point for the implementation phase.

9.7 TASK 5.6: CREATE THE USER MANUAL

The user manual should provide the system user with sufficient information to be able to use and control the system in an efficient manner. The manual defines the interface between the new computer system and the manual system.

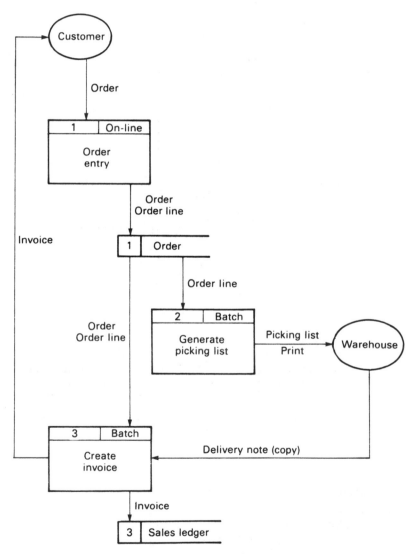

Fig. 9.10 Level 1 required physical DFD

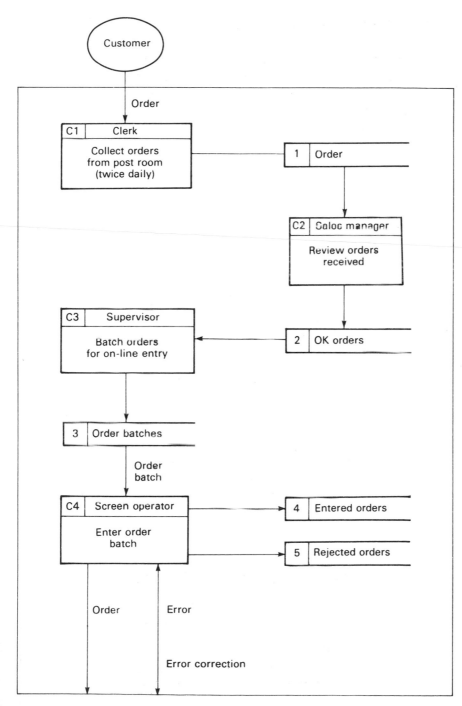

Fig. 9.11 Level 2 required physical DFD

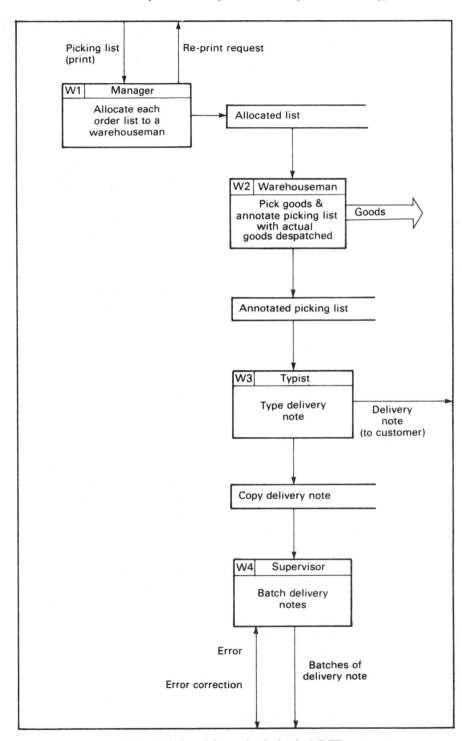

Fig. 9.12 Level 2 required physical DFD

Data flow diagrams are one way of describing activities external to the computer system. Figure 9.10 shows a level 1 required physical DFD. The external entities may be exploded into their own level 2 DFDs, each diagram using the standard DFD conventions of external entity, data store, data flow and, in this case, manual function.

Figures 9.11 and 9.12 provide examples of level 2 DFDs for the customer external entity and the warehouse external entity.

Figure 9.11 describes the method of input for orders received from customers. The sequence of functions is easily identified from the diagram. Each function now needs to be described in detail if it is to be included in the user manual. In the example, function C2, the review of orders received by the sales manager, could well be excluded from the user manual. Function C4 is the only function which directly interacts with the computer system and is, therefore, the only one which must be described in the user manual.

The user manual must describe the procedure for the input of orders, for example, the use of menus and the choice of screen format. It must also describe the error messages and the methods of error correction. Further, in this example, since the input of orders is closely related to process C3, batch orders for on-line data entry, then the user manual should describe batching and control procedures.

Many installations have their own format for user manuals. In these cases, DFDs may be used as a technique to assist in the completion of the user manual.

Figure 9.12 describes the user processing of an output report, in this case the picking list. Function W1 is the first to receive output from the computer; procedures for requesting a re-run of the print generation programs should be described, in the event of error or loss. Note also that function W4 involves input to the computer system. This function again requires careful description, this time describing the procedures for batch input and error correction.

The topics covered by the user manual described using DFDs cover the input and output procedures. There are many other areas of interest to the user. These should be described in further sections within the user manual and include:

• the days and times when the programs should be available
• procedures for start-up and close-down of the system, for example logging on and off via password identification

- procedures for back-up in the event of system failure, either manual or via some other computer system
- procedures for recovery after a failure.

9.8 TASK 5.7: CREATE THE OPERATIONS MANUAL

The operations manual provides the detailed technical documentation of the operational requirements. Many of the sections are common to the user manual and the same techniques should be employed. Detailed documentation of such items as the batch run flows, the files required for on-line programs and the data control and data preparation procedures are included in the operations manual.

The format of the operations manual will again be installation dependent. Further, it may not be possible to complete all sections of the manual at this time. However, the development team should hand over to the installation team a manual completed up to an agreed point.

The operating schedule and requirements are the prime documents completed by this task. Typically, for each on-line program, the manual should include:

- the days and times when the programs should be available
- the set-up and the close-down requirements for the programs
- the resource requirements in terms of direct access space and memory space, including the files or database areas required
- the instructions covering any non-standard processing, e.g. recovery from a system failure.

9.9 STAGE C SUMMARY

The final output from the stage is:
- a physical files or database specification, including the data dictionary
- a set of program specifications
- an implementation plan
- an operations manual
- a user manual
- the system documentation produced by all stages of structured systems analysis and design.

The completion of the physical design concludes the work of the analysts and designers. The project is now handed over to an implementation team which might have no common membership with the team of analysts and designers. The completion of stage C represents, therefore, a major milestone in the project.

The implementation team will wish to study the system specification in great detail. An initial walkthrough of the specification with representatives of both teams provides an ideal vehicle for education of the implementation team and for a detailed review of the specification.

Chapter 10
Information Systems Methodologies

10.1 THIS TEXT

This text has described a structured approach to information systems development. It has introduced a number of key techniques and fitted them into a disciplined approach based upon phases, stages, parts and tasks. The text has not attempted to follow any one methodology exactly; this is difficult because of the continuous development of commercially available methodologies. The principles behind the techniques and the understanding of a structured disciplined approach will enable the reader to understand commercially available methodologies.

This chapter includes a very short introduction to SSADM and LBMS Systems Engineering to introduce the reader to the methods. The summaries of the methods represent the author's best understanding, which may not correspond exactly with the method suppliers. The reader is advised to contact the suppliers for the latest descriptions.

10.2 FURTHER READING

Two valuable texts concerned with information systems are *Information Systems Methodologies, A Framework for Understanding*, T. William Olle *et al.*, and *Information Systems Development, Methodologies, Techniques and Tools*, D.E. Avison and G. Fitzgerald.

Information Systems Methodologies is written with a perspective free of any methodology. The text takes a comparative approach covering the many different ways of analysing and designing an application. The text assumes prior knowledge of at least one information systems methodology. It defines a framework into which many existing methodologies can be fitted to gain a new perspective on the subject of information systems methodologies.

Information Systems Development explains some of the more recent methodologies, techniques and tools. The objective is to equip the reader to understand the principles involved, to be able to judge the benefits claimed and to be able to contribute better to information systems development. The text attempts to classify methodologies and to suggest why each class is important and appropriate. The overall purpose is to bring some order to the methodology jungle.

10.3 DEVELOPING A METHODOLOGY

Many organisations have recently fully adopted commercially available methodologies; many others have developed their own methodology using the techniques and methodological notions to build a tailored structured approach. One of the latter is Illingworth Morris Information Services Ltd. They examined SSADM, using a series of pilot projects, in order to develop their own interpretation and methodology. Their structured approach is to divide the methodology into the business system design and the computer system design. The key steps in each stage are:

1.1 Interview and investigate.
1.2 Sketch the document flow diagram and develop the entity model.
1.3 Sketch the data flow diagrams and a route map.
1.4 Draw up the problems and requirement list.
1.5 Cross check the DFDs and route map against the document flow diagram and the entity model, updating the problems and requirements list as appropriate.
1.6 Present the documentation to the users.
1.7 Present the documentation to the development team.
2.1 Normalise the user documents and required output.
2.2 Draw up the third normal form relations and compare them with the entity model to create the logical entity model.
2.3 Draw up an event/entity matrix.
2.4 Produce the entity life histories for the volatile entities.
2.5 Review any changes to the problems and requirements list.
2.6 Produce the logical access maps.
2.7 Produce the event catalogue.
2.8 Produce a catalogue of common processing.
2.9 Review and approve the documentation.

2.10 Set up the database schema on the target system.
2.11 Produce an action diagram for each program.
2.12 Review and approve the system before programming.
The methodology is under continuous review and development itself. The techniques are those from the text and the approach is structured and disciplined.

10.4 SSADM

Version 3.5 of SSADM comprises six stages: systems analysis, stages 1 and 2; business systems design, stages 3, 4 and 5; and physical design, stage 6.

Stage 1, analysis of the current system, produces the current physical DFDs, a current entity model and a problems and requirements list. Stage 2, the specification of requirements, produces the current logical DFDs. These are refined, together with the stage 1 documentation, by examination of the business options, to produce the required DFDs, the required entity model, the entity life histories, an event catalogue and the dialogue specification.

Business system design includes selection of a technical solution for the specification and production of a detailed logical design. The system design stages produce the detailed logical data design and the logical update and enquiry processing.

Physical design is the conversion of the logical design into a design which fits the target hardware and software.

Schematics of version 4 of SSADM are shown in Appendix B.

10.5 LBMS SYSTEMS ENGINEERING

Learmonth and Burchett Management Systems announced their major development of the LBMS Structured Development Method in 1990. The method is known as LBMS Systems Engineering. It comprises six stages:

Stage PI   Project Initiation
Stage RA   Requirements Analysis
Stage LD   Logical Design
Stage PD   Physical Design

Stage CO   Construction
Stage IN   Installation.

Each stage comprises a number of parts and tasks. Stage PI, shown in Figure 10.1, is concerned with scoping, assessing feasibility and planning the project. Stage RA, shown in Figure 10.2, comprises two parts, an optional analysis of the current system and a specification of requirements. Note that the techniques entity or data modelling, data flow diagramming, dialogue design and prototyping are used in this stage. Stage LD, shown in Figure 10.3, uses the technique of normalisation or relational data analysis and entity life histories to produce a logical design. Stage PD, shown in Figure 10.4, designs and tunes the technical aspects of the data and the processing Stage CO, shown in figure 10.5, concerns programming, program testing, unit and link testing and system testing. Stage IN, shown in Figure 10.6, concerns installation and hand-over.

Stage P1: Project initiation

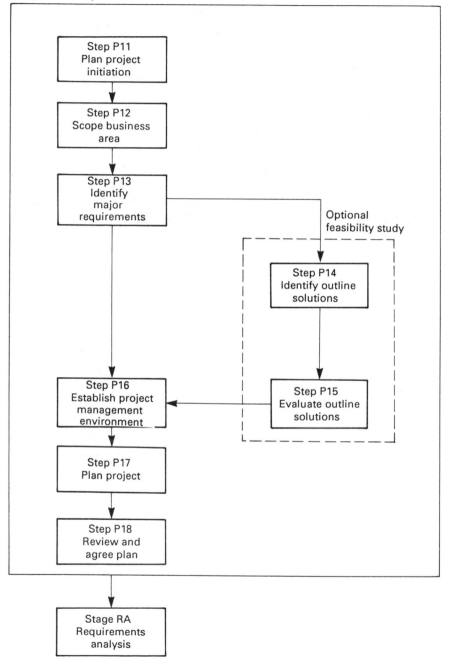

Fig. 10.1

Stage RA: Requirements analysis

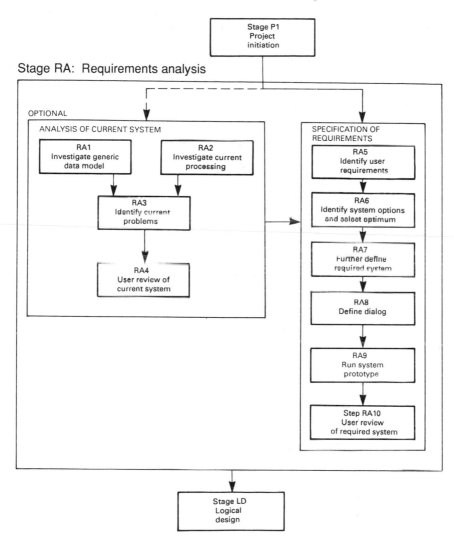

Fig. 10.2

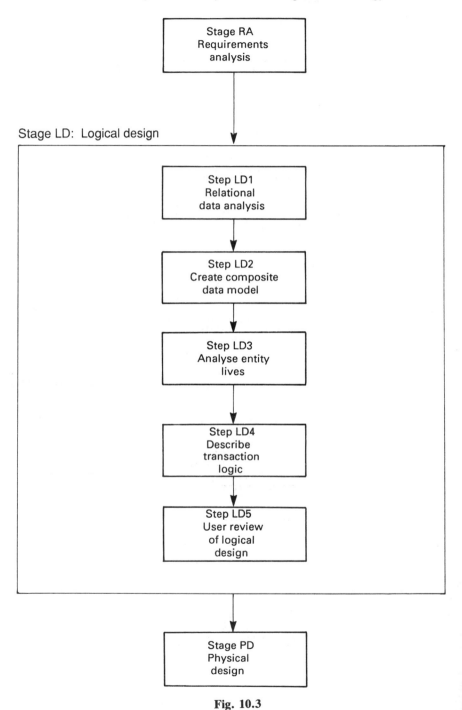

Fig. 10.3

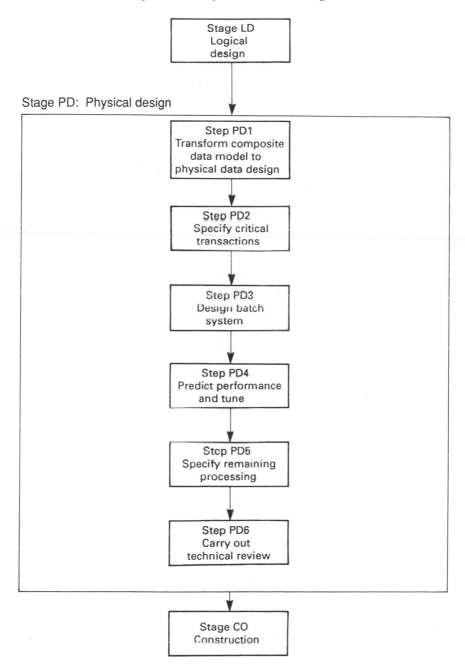

Fig. 10.4

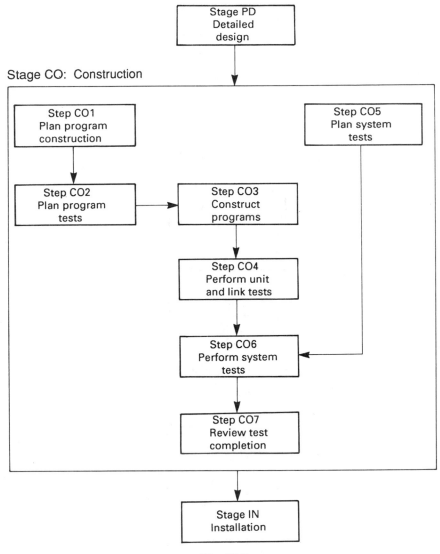

Fig. 10.5

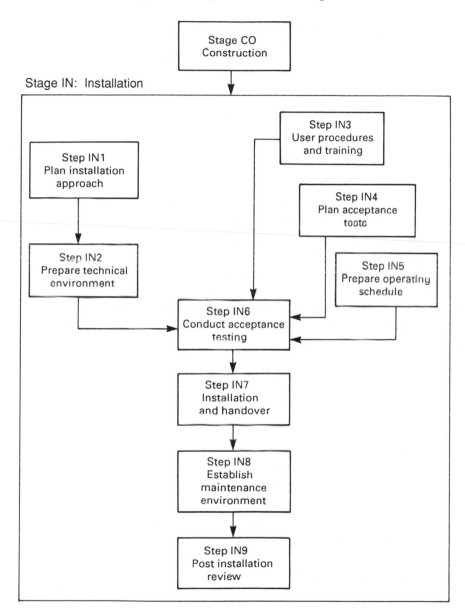

Fig. 10.6

10.6 METHODOLOGIES FOR THE 1990s

The 1980s saw a major development period for methodologies; the 1990s will see the introduction of new methodologies such as the object oriented approach and reverse engineering.

Maintenance still accounts for up to 70% of the effort expended by many computer professionals. As systems become older new people have to take over and maintain systems. Without proper development methods and documentation it is often impossible to understand the system. Reverse engineering aims to reverse the systems development cycle to obtain a specification from the code and the limited documentation available. The Institution for Electrical and Electronic Engineer's Software Journal, January 1990, was devoted to software maintenance and reverse engineering.

The object oriented approach aims to concentrate on business goals and not on the separation of understanding into data and processing. Many consultants believe that the way forward is via a higher level of abstraction than DFDs and entity models to the identification and implementation of objects.

10.7 CONCLUSIONS

The 1980s experienced a very rapid growth of Information Systems Methodologies; this rapid development will continue on into the 1990s. The majority of students at polytechnics and universities are being taught a structured approach; these people will enter employment in the 1990s. Many organisations are experiencing the benefits of a structured disciplined approach, many use techniques and most are starting to experience the benefits. The development and adoption of structured methods will continue.

Further, the 1980s saw the arrival of CASE products such as Auto-Mate Plus and Systems Engineer. These products support the majority of the techniques used in the methodologies from diagram editors to database and code generation. The combination of structured methodology and CASE tool is a powerful way to develop systems.

PART 3

THE CASE STUDY

Chapter 11
SKC Part 1: Analysis of the Current System

11.1 INTRODUCTION

The Silent Kitchen Company manufactures complete dishes, such as coq-au-vin, which are frozen for sale to restaurants. The restaurant simply places the meal in a microwave oven prior to serving. In this way, even small restaurants can offer a large à la carte menu at reasonable prices.

Customers of the Silent Kitchen Company (SKC) send their orders to the sales department. Orders are acknowledged and transcribed on to pink order forms. Each order may be for one or more dishes. Customers may also enter into contracts with sales. A contract is for one dish only, and offers it at a discount for an undertaking by the customer to purchase a fixed quantity over a fixed period. Customers may have many contracts. Orders may, therefore, contain some dishes quoting a contract price. A copy of all contract forms are sent to accounts for invoice pricing. All other invoices are priced from a price list prepared by the kitchen staff. Sales receive a daily stock list of the dishes available for sale from the kitchen.

Accounts perform the invoicing and sales ledger functions. Invoices are sent to customers when a copy of the delivery note is received from the kitchen. Accounts also prepare credit notes for returned dishes and provide, for sales, a weekly stop list of customers who have exceeded their credit limit. Accounts vet and authorise new customers and send details about new customers to sales so that they may update their wall chart. Accounts send statements to customers and record payments received from customers.

The kitchen, apart from performing the obvious function, deals with pink order forms, the daily stock list, delivery notes and returned dishes. For each dish returned by a customer, the kitchen staff prepare a returned dish notification form for accounts. When a delivery of ingredients is received, the kitchen staff send an ingredients received note to accounts and prepare an ingredients received list for purchasing. This is to allow purchasing to check the invoice received from the supplier against the actual ingredients received. When any ingredient is running short, a low ingredient card is prepared and sent to purchasing who raise a purchase order on the supplier. Purchasing handle all the purchase ledger transactions.

A computer assisted system is required initially for the sales and accounting functions. Structured systems analysis and design is to be used to design a new system.

11.2 INTERVIEW TRANSCRIPT – SALES MANAGER RESPONSIBLE FOR THE SALES OFFICE

The first task each morning for me is to check how many signed contracts have arrived in the post. This is my major function – negotiating contracts with customers. For each new signed contract a copy is made for accounts and the original is filed in the contracts file.

When orders are received quoting a contract price, the contracts clerk retrieves the contract from the file, records on it the quantity ordered and returns it to the file. Each contract is negotiated for one dish only; an order, therefore, may reference several contracts if it is for more than one dish.

All orders are checked for credit worthiness against the weekly stop list by a junior clerk. They are then placed on an orders outstanding pile.

The manageress, apart from looking after the typing pool, is responsible for checking whether dishes are available to meet the outstanding orders. All pending orders are checked first each day against the stock list. If there is sufficient stock for all the dishes on the order, then the order is transcribed on to a number of pink order forms. The pink order forms create the dishes or order lines for despatch pile. If there is insufficient stock, then the *complete* order is returned to the pending orders pile. When all pending orders have been checked, today's orders are checked in the same fashion.

Each day then, a number of order lines are placed in the ready-to-despatch pile. The typing pool then prepare an order acknowledgment for the customer and a typed version of the pink order form for the kitchen. Names and addresses are extracted from a name and address wall chart which is kept up to date by a junior clerk from information supplied by accounts.

11.3 INTERVIEW TRANSCRIPT – CHIEF ACCOUNTANT RESPONSIBLE FOR ACCOUNTING

My two major functions as chief accountant are the authorisation of new customer accounts and the authorisation of credit notes.

When a potential customer applies for an account, they are vetted for credit worthiness. If satisfactory references are received, their name and address and credit status are added to the sales ledger and notification of a new customer is sent to sales.

Credit note authorisations result from the return of dishes for a variety of reasons. I raise a credit authorisation slip which goes to a typist who types a credit note. The credit note itself is sent to the customer and the credit is recorded on the sales ledger.

We also receive copies of delivery notes from the kitchen. For each delivery note, an invoice is typed and sent to the customer. Full details of the invoice are typed on to a ledger sheet. Payments are recorded on the ledger sheets by a clerk. We keep one sheet per customer; these sheets comprise the sales ledger.

Invoice pricing is tricky. Two files are needed: a standard price list for standard orders and the contracts file for orders against a contract. The standard price list is provided weekly by the kitchen staff.

Every week each sales ledger sheet is examined to identify customers who have over-stepped their credit limit. A stop list is compiled and sent to sales. Finally, at the end of every month, all the sales ledger sheets are photocopied. The copies are sent to our customers in the form of sales statements.

The kitchen staff send us ingredient received notes but we have no use for them.

Investigation

These notes and interview transcripts make up part of the output from part 1 task 1.1, investigation.

At this early stage in the project, a brief introduction and two interviews, the understanding of SKC's current system can only be at a general level. The exact detail of how some functions are carried out has not been investigated or documented.

It is possible, from the investigation notes, to construct the document flow diagram, a level 1 current physical DFD, level 2 current physical DFDs for sales and accounting, and the entity model. It is necessary to make some assumptions. List your assumptions while constructing the diagrams, then compare your diagrams which those provided to ascertain if similar assumptions were made.

11.4 SKC: CURRENT PHYSICAL DATA FLOW DIAGRAMS, TASK 1.2

Figure 11.1 shows the document flow diagram, Figure 11.2 shows the level 1 DFD, Figure 11.3 shows the level 2 DFD for sales, and Figure 11.4 shows the level 2 DFD for accounts.

DATA FLOW DIAGRAM

SYSTEM: *SKC - SALES & ACCOUNTING* DATE:

AUTHOR: *G. CUTTS* PAGE: / of /

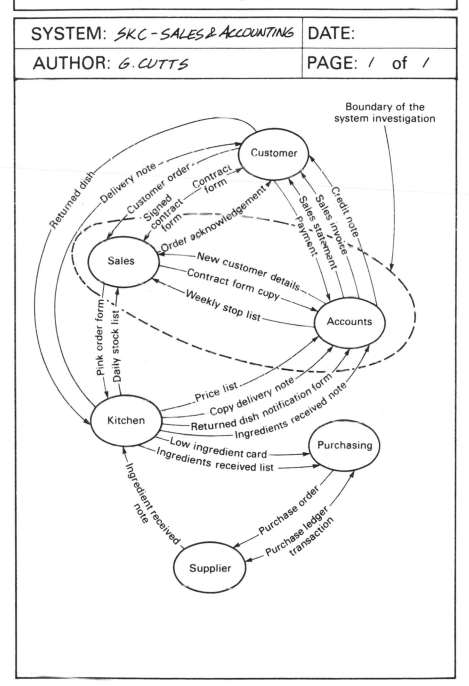

Fig. 11.1

DATA FLOW DIAGRAM

SYSTEM: *SKC – SALES & ACCOUNTING*	DATE:
AUTHOR: *G. CUTTS*	PAGE: *1* of *3*

LEVEL: *1*	CURRENT/~~REQ~~.	PHYS./~~LOGICAL~~.

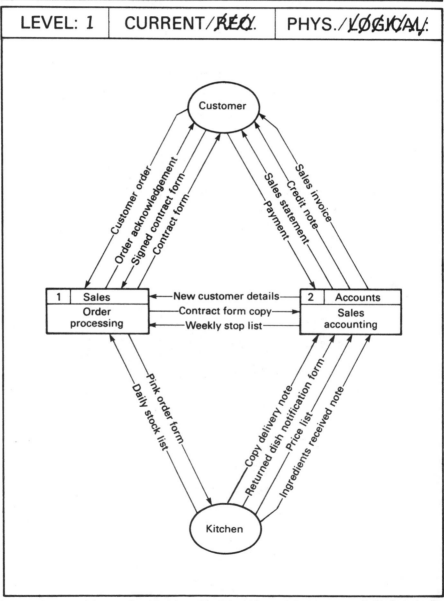

Fig. 11.2

DATA FLOW DIAGRAM

SYSTEM: *SKC – SALES & ACCOUNTING* | DATE:

AUTHOR: *G. CUTTS* | PAGE: *2* of *3*

LEVEL: *2* | CURRENT/~~REQ~~. | PHYS./~~LOGICAL~~.

TITLE: *SALES*

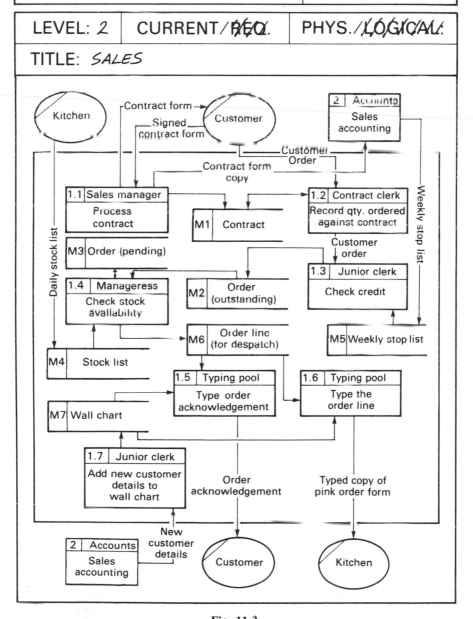

Fig. 11.3

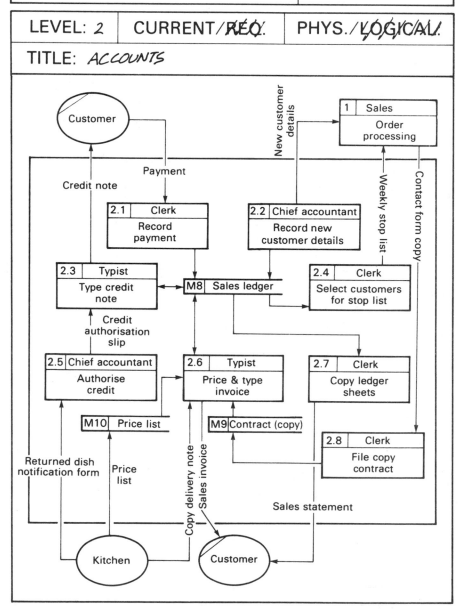

DATA FLOW DIAGRAM

| SYSTEM: *SKC-SALES & ACCOUNTING* | DATE: |
| AUTHOR: *G. CUTTS* | PAGE: *3* of *3* |

| LEVEL: *2* | CURRENT/~~REQ~~. | PHYS./~~LOGICAL~~ |
| TITLE: *ACCOUNTS* | | |

Fig. 11.4

11.5 SKC: CURRENT ENTITY MODEL, TASK 1.3

Figure 11.5 shows the entity matrix and Figure 11.6 shows the current entity model.

The order within the SKC system may exist as an outstanding order, a pending order or an order for despatch. It is necessary to be able to access the orders by this status. An operation owner, order status, is shown on the entity model.

	Customer	Contract	Order head	Order line	Dish	Invoice	Payment	Credit note
Customer		*	*			*	*	*
Contract				*				
Order head				*				
Order line								
Dish		*		*				
Invoice								
Payment								
Credit note								

Fig. 11.5

ENTITY MODEL

SYSTEM: SKC. - SALES & ACCOUNTING	DATE:
AUTHOR: G. CUTTS	PAGE: 1 of 1

VERSION: SYSTEMS ANALYSIS

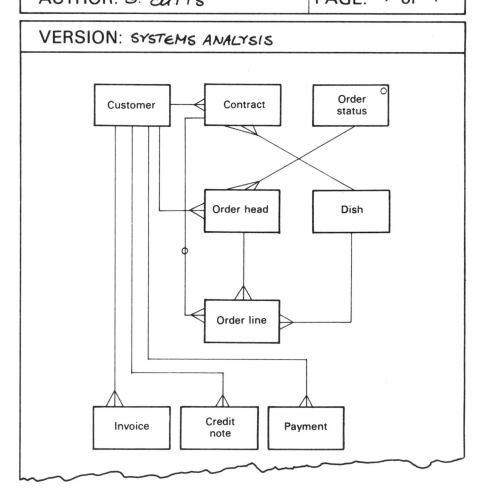

Fig. 11.6

11.6 SKC: DATA STORE ENTITY CROSS REFERENCES, TASK 1.4

The physical data store entity cross reference (Figure 11.7) indicates for each data store the entities that are within the data store. Figure 11.8 shows which entities are accessed by each function within each data store.

Four logical data stores were identified and are shown in Figure 11.9. Data stores L3, product, and L4, sales ledger, appear very straightforward, as also do data stores for customer and order head/order line. This leaves one question – where to locate the contract entity? Possibilities are: with the customer entity, with the dish entity, or with neither.

For this system customer is chosen, as the contract entity is most used, on-line, for order processing (contract or standard order?) and invoicing (contract or standard price?).

Four logical data stores seem appropriate. These are represented in Figure 11.10.

DATA STORE/ENTITY X REF.

SYSTEM: SKC – SALES & ACCOUNTING

DATE:

AUTHOR: G. CUTTS

PAGE: 1 of 1

PHYSICAL/~~LOGICAL~~:

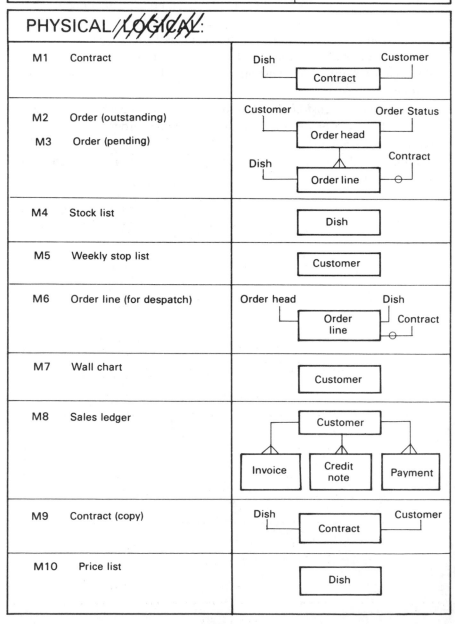

M1 Contract	Dish — Customer, Contract
M2 Order (outstanding) M3 Order (pending)	Customer — Order Status, Order head; Dish — Order line — Contract
M4 Stock list	Dish
M5 Weekly stop list	Customer
M6 Order line (for despatch)	Order head — Dish, Order line — Contract
M7 Wall chart	Customer
M8 Sales ledger	Customer — Invoice, Credit note, Payment
M9 Contract (copy)	Dish — Customer, Contract
M10 Price list	Dish

Fig. 11.7

DATA FLOW DIAGRAM

SYSTEM: *SKC- SALES & ACCOUNTING*	DATE:
AUTHOR: *G. CUTTS*	PAGE: *2* of *3*

LEVEL: *2*	CURRENT/R̶E̶Q̶/	PHYS./L̶O̶G̶I̶C̶A̶L̶/

TITLE: *SALES*

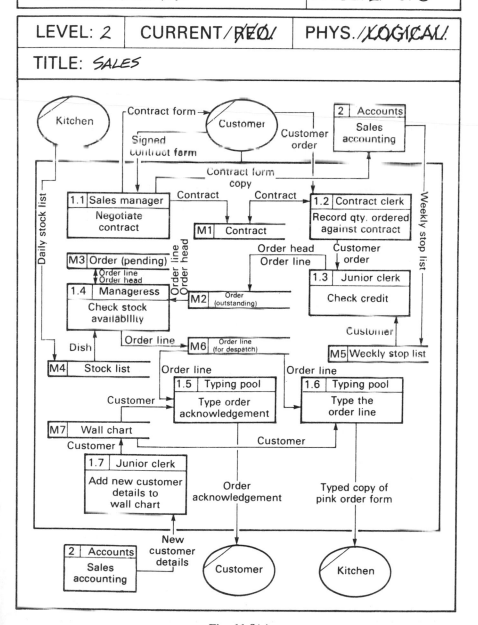

Fig. 11.8(a)

DATA FLOW DIAGRAM

SYSTEM: *SKC-SALES & ACCOUNTING*	DATE:
AUTHOR: *G. CUTTS*	PAGE: *3* of *3*

LEVEL: *2*	CURRENT/~~REQ.~~	PHYS./~~LOGICAL~~

TITLE: *ACCOUNTS*

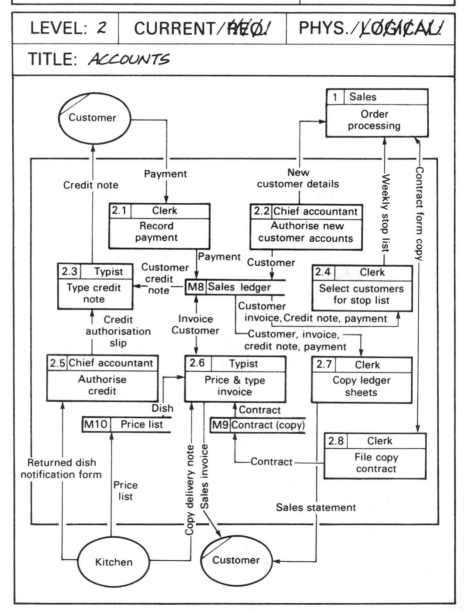

Fig. 11.8(b)

ENTITY MODEL

SYSTEM: SKC - SALES & ACCOUNTING	DATE:
AUTHOR: G. CUTTS	PAGE: 1 of 1

VERSION: SYSTEMS ANALYSIS

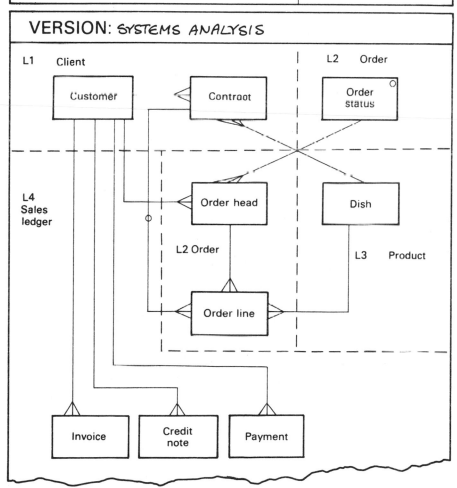

Fig. 11.9

DATA STORE/ENTITY X REF.

SYSTEM: SKC - *SALES &*
ACCOUNTING   DATE:

AUTHOR: G. CUTTS   PAGE: 1 of 1

PHYSICAL/LOGICAL:

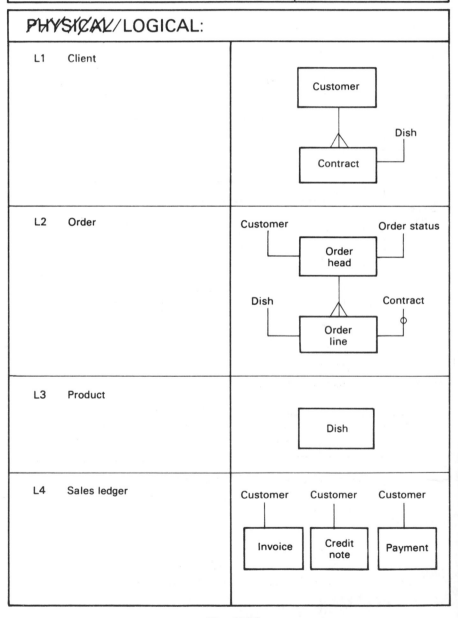

L1	Client
L2	Order
L3	Product
L4	Sales ledger

Fig. 11.10

11.7 SKC: CURRENT LOGICAL DATA FLOW DIAGRAMS, TASK 1.5

11.7.1 Step 1: Logicalise the data flows

The following data flows have been renamed and are shown in Figure 11.11:

From (physical)	To (logical)
contract form	customer contract
signed contract form	customer contract
contract form copy	customer contract
daily stock list	dish stock
weekly stop list	credit status
new customer details	customer
pink order form	despatch note
copy delivery note	delivery note
returned dish notification form	dish reference and quantity
credit authorisation slip	credit authorisation

11.7.2 Step 2: Delete time dependencies

The following functions and data stores have been marked for deletion in Figure 11.11:
Function 1.7, zero logical function
Function 2.8, zero logical function
Data stores M2 and M6, physical time dependency, logically not required.

11.7.3 Step 3: Logicalise the functions

The following functions have been renamed or deleted:

1.5 type order acknowledgement	renamed produce order acknow-ledgement
1.6 type the order line	renamed produce despatch note
1.7 add new customer details	deleted
2.3 type credit note	renamed produce credit note
2.6 price and type invoice	renamed price and produce in-voice
2.7 copy ledger sheets	renamed produce sales statement
2.8 file copy contract	deleted

Figure 11.11 shows the level 2 current physical DFDs processed by the first three steps.

DATA FLOW DIAGRAM

SYSTEM: *SKC- SALES & ACCOUNTING*	DATE:
AUTHOR: *G. CUTTS*	PAGE: *2* of *3*

LEVEL: *2*	CURRENT/~~REQ~~	PHYS./~~LOGICAL~~

TITLE: *SALES*

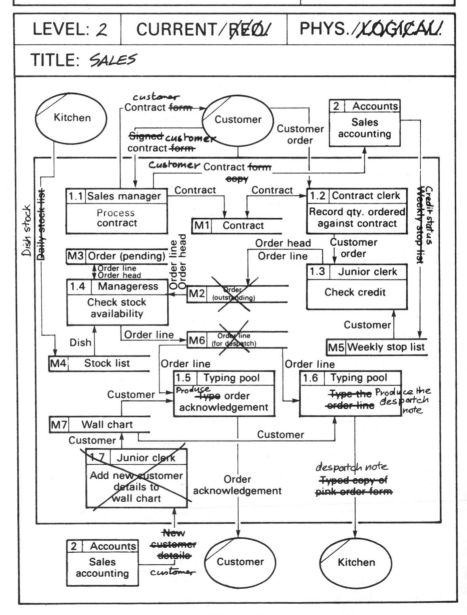

Fig. 11.11(a)

DATA FLOW DIAGRAM

SYSTEM: *SKC-SALES & ACCOUNTING*	DATE:
AUTHOR: *G. CUTTS*	PAGE: *3* of *3*

LEVEL: *2*	CURRENT/~~REQ~~	PHYS./~~LOGICAL~~

TITLE: *ACCOUNTS*

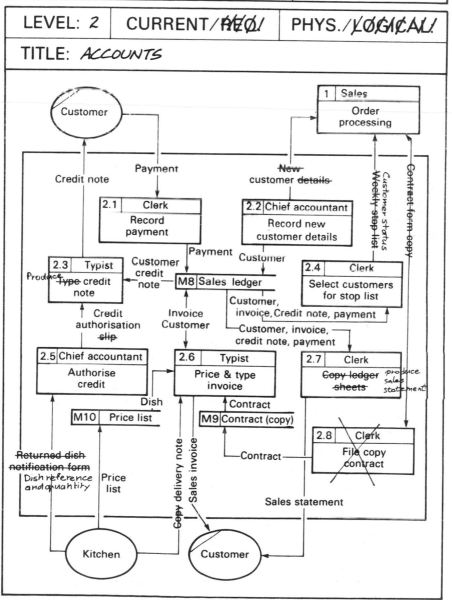

Fig. 11.11(b)

11.7.4 Step 4: Create the current logical DFDs

Figures 11.12 and 11.13 show the current logical DFDs. Logicalised data flows and functions were copied to the current logical DFDs, function by function. Where a data flow accessed a data store, the physical data store was replaced by its logical equivalent.

The action taken on transfer of each function is given below:

Function

1.1 Contract is written to data store L1 for access by function 1.2 and accounts. The identical logical data flow to accounts is deleted.

1.2 The customer order represented by the entities order head and order line is inserted into data store L2.

1.3 Physical data store M5 is replaced by logical data store L1. The status of the order is modified; access is not required to the order line.

1.4 M4 replaced by L3. All other transactions are performed by status manipulation.

1.5 M7 replaced by L1 and M6 by L2.

1.6 M6 replaced by L2.

1.7 A new function is required to update the quantity in stock attribute in the dish entity.

2.1 M8 replaced by L4.

2.2 The entity inserted by this function, originally physical data store M8 with a copy of the customer entity on M7, is now inserted into logical data store L1. Sales have access to the data store.

2.3 M8 replaced by L4.

2.4 List of customer numbers replaced by status on L1. M8 replaced by L4.

2.5 No change.

2.6 M10 replaced by L3, M9 replaced by L1, and M8 replaced by L4 and L1.

2.7 M8 replaced by L4 and L1.

Figure 11.13 was then constructed from Figure 11.12. It shows the level 1 current logical DFD.

11.8 SKC: PROBLEMS AND REQUIREMENTS LIST, TASK 1.6

The problems and requirements list was developed during the previous tasks and is now formally documented in Figure 11.14.

DATA FLOW DIAGRAM

SYSTEM: *SKC SALES & ACCOUNTING*	DATE:
AUTHOR: *G. CUTTS*	PAGE: *2* of *3*

LEVEL: *2*	CURRENT/~~REQ~~.	~~PHYS~~./LOGICAL:

TITLE: *SALES*

Fig. 11.12(a)

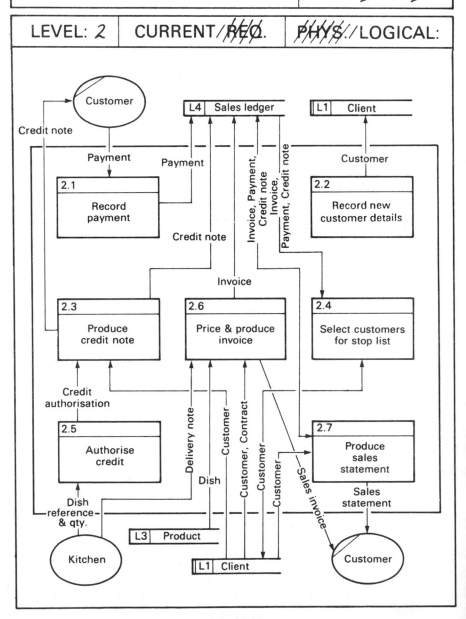

Fig. 11.12(b)

DATA FLOW DIAGRAM

SYSTEM: *SKC-SALES & ACCOUNTING*	DATE:
AUTHOR: *G. CUTTS*	PAGE: *1* of *3*

LEVEL: *1*	CURRENT/R̶E̶Q̶.	P̶H̶Y̶S̶./LOGICAL:

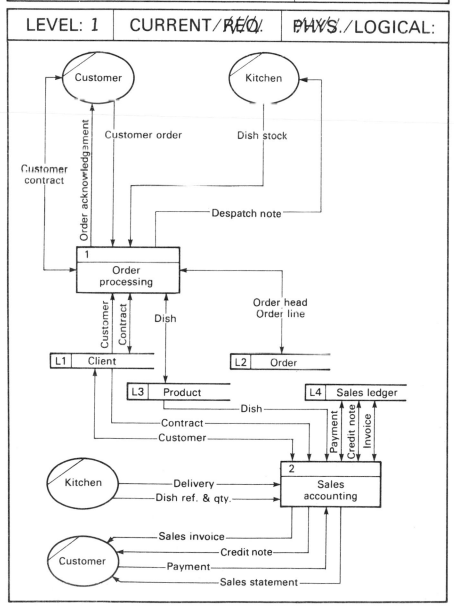

Fig. 11.13

PROBLEMS/REQUIREMENTS LIST

SYSTEM: *SKC - SALES & ACCOUNTING*	DATE:
AUTHOR: *G. CUTTS*	PAGE: *1* of *1*

No.	Problem/requirement	Init.	Solution Reference
1.	Copies of contracts are not always sent to accounting. This results in incorrect pricing since orders for contract items may be priced at the standard higher price. This in turn causes customer problems.	C.A.	
2.	Three files are used for processing orders: outstanding order, pending order and order lines for despatch. This results in considerable movement of documents, loss of efficiency and lost documents. Additionally, no order can be processed until all the dishes on the order are in stock.	S.M.	
3.	There is no easy method of selecting pending orders for despatch when new stock is delivered. Many pending orders are processed each day for several weeks especially for long lead times on delivery of new stock.	S.M.	
4.	Management require a report on order throughput. Some information can be provided but other information such as average time from receipt of an order to despatch is difficult or impossible to provide.	M.D.	
5.	Invoice typing is several days behind despatch.	S.A.	
6.	A statement for customers in the current form, photocopied ledger card is not acceptable.	C.A.	
7.	The process "select slow payers" is not done. It is too complicated to summarise the ledger sheets.	C.A.	
8.	Invoices are currently raised for every delivery. It is required to invoice weekly with invoices listing all deliveries for the week.	C.A.	
9.	It is required to part deliver orders. Instead of complete orders being held pending, only order lines should be held.	S.M.	
10.	The current brought forward ledger system should be converted to an open item ledger. Credit notes and payments must be against specific invoices not on account.	C.A.	

Fig. 11.14

Chapter 12
SKC Part 2: Specification of Requirements

This chapter describes the second part of structured systems analysis and design applied to SKC. A full specification of requirements is produced.

12.1 SKC: REQUIRED LOGICAL DATA FLOW DIAGRAMS, TASK 2.1

Figure 12.1 shows the level 1 required logical DFD; Figure 12.2 (a) and (b) shows the level 2 required logical DFDs; and Figure 12.3 restates the problems and requirements list with solution references.

DATA FLOW DIAGRAM

SYSTEM: *SKC –SALES 2ACCOUNTING*	DATE:
AUTHOR: *G.CUTTS*	PAGE: / of **4**

LEVEL:	~~CURRENT~~/REQ.	~~PHYS.~~/LOGICAL:

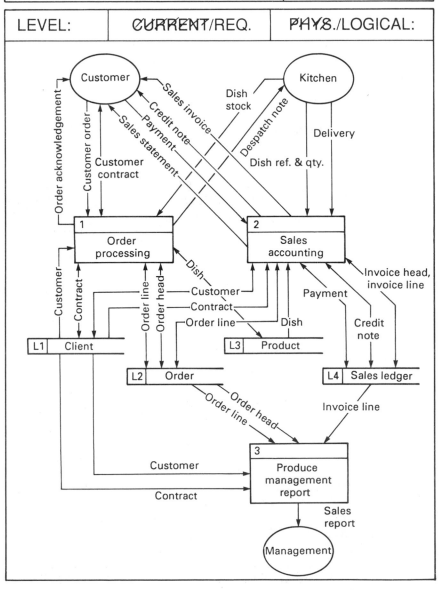

Fig. 12.1

DATA FLOW DIAGRAM

SYSTEM: *SKC – SALES & ACCOUNTING*	DATE:
AUTHOR: *G. CUTTS*	PAGE: *2* of *3*

LEVEL: *2*	~~CURRENT~~/REQ.	~~PHYS.~~/LOGICAL:

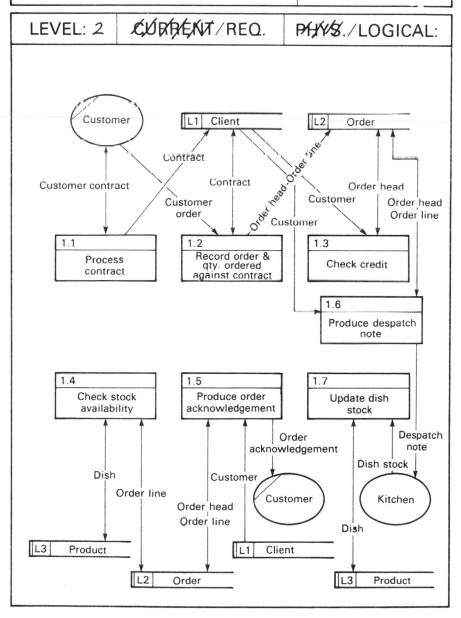

Fig. 12.2(a)

DATA FLOW DIAGRAM

SYSTEM: *SKC - SALES & ACCOUNTING*	DATE:
AUTHOR: *G. CUTTS*	PAGE: *3* of *3*

LEVEL: *2*	C̶U̶R̶R̶E̶N̶T̶/REQ.	P̶H̶Y̶S̶./LOGICAL:

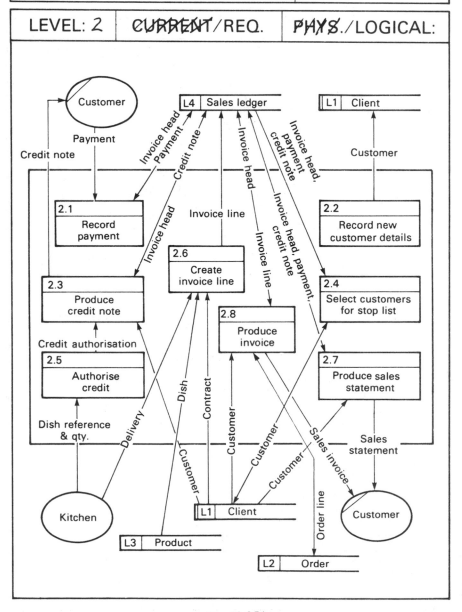

Fig. 12.2(b)

PROBLEMS/REQUIREMENTS LIST

SYSTEM: *SKC – SALES & ACCOUNTING*   DATE:

AUTHOR: *G. CUTTS*   PAGE: *1* of *1*

No.	Problem/requirement	Init.	Solution Reference
1.	Copies of contracts are not always sent to accounting. This results in incorrect pricing since orders for contract items may be priced at the standard higher price. This in turn causes customer problems.	C.A.	S1
2.	Three files are used for processing orders: outstanding order, pending order and order lines for despatch. This results in considerable movement of documents, loss of efficiency and lost documents. Additionally, no order can be processed until all the dishes on the order are in stock.	S.M.	S2
3.	There is no easy method of selecting pending orders for despatch when new stock is delivered. Many pending orders are processed each day for several weeks especially for long lead times on delivery of new stock.	S.M.	S2
4.	Management require a report on order throughput. Some information can be provided but other information such as average time from receipt of an order to despatch is difficult or impossible to provide.	M.D.	S3
5.	Invoice typing is several days behind despatch.	S.A.	S4
6.	A statement for customers in the current form, photocopied ledger card is not acceptable.	C.A.	S4
7.	The process "select slow payers" is not done. It is too complicated to summarise the ledger sheets.	C.A.	S4
8.	Invoices are currently raised for every delivery. It is required to invoice weekly with invoices listing all deliveries for the week.	C.A.	S5
9.	It is required to part deliver orders. Instead of complete orders being held pending, only order lines should be held.	S.M.	S2
10.	The current brought forward ledger system should be converted to an open item ledger. Credit notes and payments must be against specific invoices not on account.	C.A.	S6

Fig. 12.3

12.2 SKC: REQUIRED ENTITY MODEL, TASK 2.2

The required entity model is shown in Figure 12.4.

Each problem or requirement, Figure 12.3, has a solution reference. The references are expanded below:

S1 A common contracts data store to both sales and accounting will resolve this problem.

S2 A single orders data store is proposed with order lines recorded as separate entities. Each order line entity may take the status outstanding, pending or despatched. All order lines will be processed on each stock allocation run. These measures should overcome problems 2, 3 and 9.

S3 A new function 3, produce management report, is included on the level 1 DFD.

S4 Problems 5, 6 and 7 are relieved by automation of the processing involved.

S5 The original function 2.6, produce an invoice for each despatch, is amended to produce an invoice line on logical data store L4 for each despatch. A new function, 2.8, will produce invoices weekly for all invoice lines inserted during the previous week.

S6 Functions 2.1 and 2.3 are amended to record payments and credit notes against invoices not against the customer.

Solutions 2 and 6 require modifications to be made to the entity model. The order status has been deleted and replaced by order line status. This is to satisfy requirement 9. Outstanding, pending and despatched status should be applied to order lines, thus enabling part delivery of an order.

Requirement 10 is for an open item sales ledger. A brought forward ledger simply accounts for the total debt and the total credit; it does not record individual credit notes or payments against invoices. This can cause accounting problems with bad debts or disputed invoices. An open item ledger allocates each credit note or payment to an invoice. The entity model has been revised to show customers having many invoices, with each invoice having a number of credit notes and payments.

Invoices contain many deliveries, each creating an invoice line on the invoice. Invoices, therefore, comprise an invoice head, with many invoice lines.

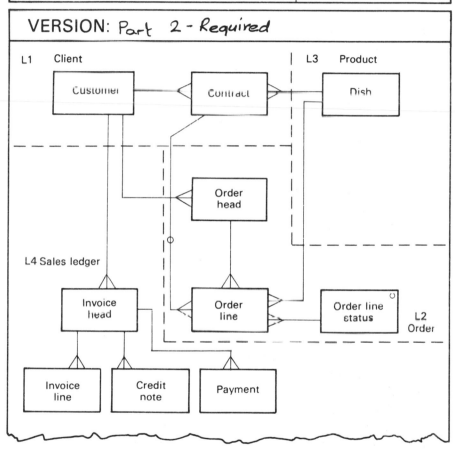

ENTITY MODEL

SYSTEM: SKC- *SALES & ACCOUNTING*   DATE:

AUTHOR: G. CUTTS   PAGE: I of I

VERSION: Part 2 - Required

L1   Client

Customer

Contract

L3   Product

Dish

Order head

L4 Sales ledger

Invoice head

Order line

Order line status

L2 Order

Invoice line

Credit note

Payment

Fig. 12.4

**12.3 SKC: OUTLINE BUSINESS SPECIFICATION,
TASKS 2.3, 2.4 AND 2.5**

For the purpose of this case study, only one option is postulated. The
level 1 and level 2 required physical DFDs are shown in Figures 12.5,
12.6 and 12.7.

The option makes maximum use of a computer to implement the
system for SKC, with functions chosen to be on-line or batch with one
exception: the function 'authorise credit' is retained as a manual
function to be carried out by the chief accountant.

DATA FLOW DIAGRAM

SYSTEM: *SKC – SALES & ACCOUNTING*	DATE:
AUTHOR: *G. CUTTS*	PAGE: *1* of *3*
LEVEL: *1* ~~CURRENT~~/REQ.	PHYS./~~LOGICAL~~:

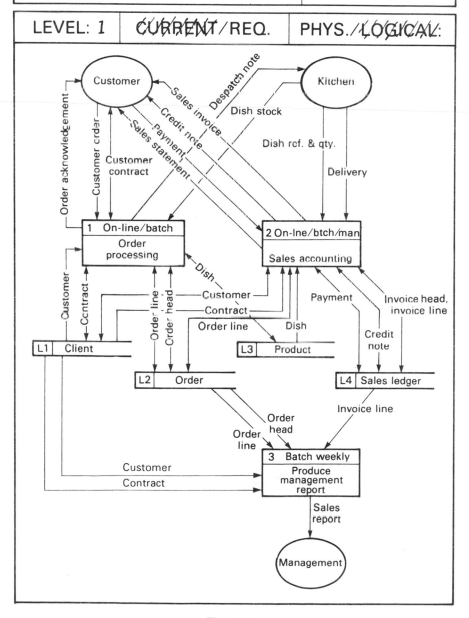

Fig. 12.5

DATA FLOW DIAGRAM

SYSTEM: *SKC SALES & ACCOUNTING*	DATE:
AUTHOR: *G. CUTTS*	PAGE: *2* of *3*

LEVEL: *1*	~~CURRENT~~/REQ.	PHYS./~~LOGICAL~~:

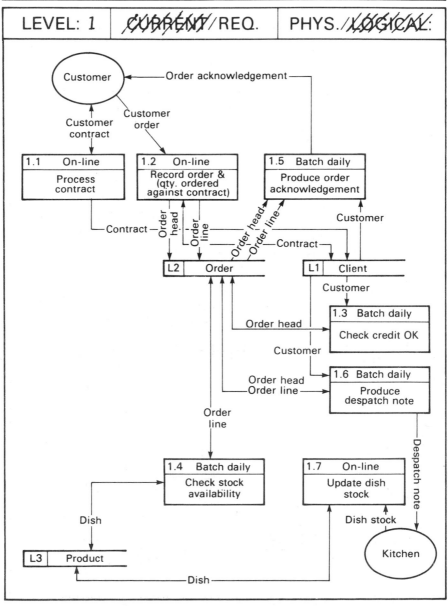

Fig. 12.6

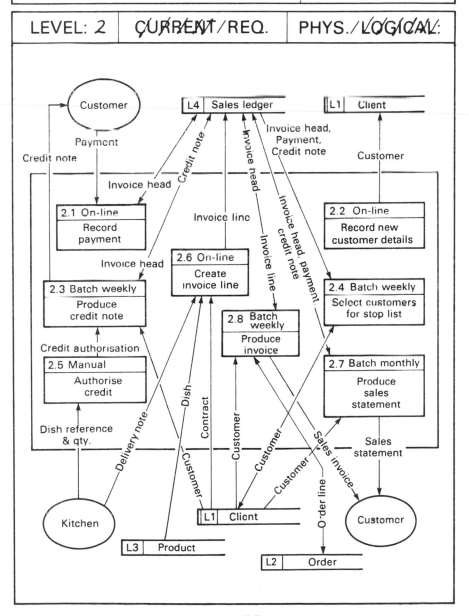

DATA FLOW DIAGRAM

SYSTEM: *SKC - SALES & ACCOUNTING*   DATE:

AUTHOR: *G. CUTTS*   PAGE: 3 of 3

LEVEL: 2   CURRENT/REQ.   PHYS./LOGICAL:

Fig. 12.7

12.4 SKC: ENTITY DESCRIPTIONS, TASK 2.6

Figures 12.8 to 12.16 document the entity descriptions for the nine entities shown on the required entity model. The entity descriptions are based on reasonable assumptions.

ENTITY DESCRIPTION

SYSTEM: SKC-SALES ACCOUNTING	DATE:
AUTHOR: G. CUTTS	PAGE: 01. of 9

NAME: CONTRACT

NARRATIVE: A contract for one dish requiring orders up to a fixed minimum quantity at a special. Contract numbers unique.

Key	Data item	Format	Len.	Comment
✓	Contract number			
	Customer number			
	Dish reference			
	Contract date			
	Expiry date			
	Contract quantity			
	Quantity ordered to date			
	Contract price			

VOLUMETRICS	
ENTITY SIZE	
No. OF OCCURRENCES	
TOTAL	

Fig. 12.8

ENTITY DESCRIPTION

SYSTEM: SKC-SALES ACCOUNTING	DATE:
AUTHOR: G. CUTTS	PAGE: 2 of 9

NAME: *CREDIT NOTE*

NARRATIVE: *A negative or inverse invoice for a returned dish. A credit note is issued against a specific invoice.*

Key	Data item	Format	Len.	Comment
✓ ✓ ✓	{{ Customer number }} {{ Invoice number }} (Credit note number) Credit note date Credit value			Credit note numbers consecutively allocated within invoice & customer

	VOLUMETRICS	
	ENTITY SIZE	
	No. OF OCCURRENCES	
	TOTAL	

Fig. 12.9

ENTITY DESCRIPTION

SYSTEM: *SKC-SALES & ACCOUNTING*	DATE:
AUTHOR: *G. CUTTS*	PAGE: *3* of *9*

NAME: *CUSTOMER*

NARRATIVE: *Customer data not related to a specific order or contract.*

Key	Data item	Format	Len.	Comment
✓	Customer number			
	Customer name			
	Customer address			5 lines
	Stop listed			yes/no
	Credit limit			£ value
	Credit time			Number of months
	Management clearance			Yes/no
	Activity			

VOLUMETRICS	
ENTITY SIZE	
No. OF OCCURRENCES	
TOTAL	

Fig. 12.10

ENTITY DESCRIPTION

SYSTEM: *SKC - SALES ACCOUNTING*	DATE:
AUTHOR: *G. CUTTS*	PAGE: *4* of *9*

NAME: *DISH*

NARRATIVE: All *data relevant to a specific dish. (not recipe).*

Key	Data item	Format	Len.	Comment
✓	Dish reference			
	Dish description			
	Number of portions in pack			
	Quantity in stock			
	Standard price			

	VOLUMETRICS	
	ENTITY SIZE	
	No. OF OCCURRENCES	
	TOTAL	

Fig. 12.11

ENTITY DESCRIPTION

SYSTEM: *SKC - SALES ACCOUNTING*	DATE:
AUTHOR: *G. CUTTS*	PAGE: *5* of *9*

NAME: *INVOICE HEAD*

NARRATIVE: *The heading of an invoice which collects together all the invoice lines for one week.*

Key	Data item	Format	Len.	Comment
✓ ✓	{ Customer number } { Invoice number } Invoice date			Invoice numbers are unique within customer

VOLUMETRICS	
ENTITY SIZE	
No. OF OCCURRENCES	
TOTAL	

Fig. 12.12

ENTITY DESCRIPTION

SYSTEM: *SKC-SALES ACCOUNTING*	DATE:
AUTHOR: *G. CUTTS*	PAGE: *6* of *9*

NAME: *INVOICE LINE*

NARRATIVE: *A line on an invoice for one delivery of one dish.*

Key	Data item	Format	Len.	Comment
✓ ✓ ✓	{ Customer number Invoice number Line number }			Invoice number added after initial insertion of the entity
	Dish reference			
	Standard price Contract price			
	Quantity delivered			
	Invoice line value			

	VOLUMETRICS	
	ENTITY SIZE	
	No. OF OCCURRENCES	
	TOTAL	

Fig. 12.13

ENTITY DESCRIPTION

SYSTEM: *SKC-SALES ACCOUNTING*	DATE:
AUTHOR: *G. CUTTS*	PAGE: *7* of *9*

NAME: *ORDER HEAD*

NARRATIVE: *Order data not related to a specific dish on an order.*

Key	Data item	Format	Len.	Comment
✓ ✓	{ Customer number { SKC order number } (Order number) Date of order (Delivery address) Order stopped			SKC order numbers are allocated within each customer Yes/No

	VOLUMETRICS	
	ENTITY SIZE	
	No. OF OCCURRENCES	
	TOTAL	

Fig. 12.14

ENTITY DESCRIPTION

SYSTEM: SKC-SALES ACCOUNTING	DATE:
AUTHOR: G. CUTTS	PAGE: 8 of 9

NAME: ORDER LINE

NARRATIVE: *The detail of an order for one dish as part of a total order.*

Key	Data item	Format	Len.	Comment
✓ ✓ ✓	{ Customer number } { SKC order number } Dish reference			
	Quantity ordered (Contract price)			
✓	Order line status			Pending/despatched

	VOLUMETRICS	
	ENTITY SIZE	
	No. OF OCCURRENCES	
	TOTAL	

Fig. 12.15

ENTITY DESCRIPTION

SYSTEM: *SKC-SALES ACCOUNTING*	DATE:
AUTHOR: *G. CUTTS*	PAGE: *9* of *9*

NAME: *PAYMENT*

NARRATIVE: *A payment of cash or cheque reducing the overall outstanding debt. Payments are recorded against a specific invoice.*

Key	Data item	Format	Len	Comment
✓ ✓ ✓	{{ Customer number }} {{ Invoice number }} { Payment number } Payment type Payment date Payment value			Payment numbers are unique within customer & invoice.

VOLUMETRICS	
ENTITY SIZE	
No. OF OCCURRENCES	
TOTAL	

Fig. 12.16

12.5 SKC: INPUT AND OUTPUT DESCRIPTIONS, TASK 2.7

Figures 12.17 to 12.25 document the input and output descriptions for the nine data flows. The input and output descriptions are based on reasonable assumptions.

INPUT OUTPUT DESCRIPTION

SYSTEM: *SKC-SALES ACCOUNTING*	DATE:
AUTHOR: *G. CUTTS*	PAGE: *1* of *9*

NAME: *CREDIT NOTE*

NARRATIVE: *A credit note is allocated against a specific invoice for a given customer.*

Key	Data item	Format	Len.	Comment
	Customer number			
	Name			
	Address			
	Credit note number			Credit note number
	Credit note date			unique within
	Credit value			invoice & customer
	Invoice number			
				Credit can only
				be given against
				a specific invoice

VOLUMETRICS

Fig. 12.17

INPUT OUTPUT DESCRIPTION

SYSTEM: *SKC-SALES ACCOUNTING*	DATE:
AUTHOR: *G. CUTTS*	PAGE: *2* of *9*

NAME: *CUSTOMER CONTRACT*	

NARRATIVE: *The content of the negotiation between a customer and SKC.*

Key	Data item	Format	Len.	Comment
	Contract number			A unique number within SKC
	Customer number			
	Name			
	Address			
	Dish reference			
	Dish description			
	Contract date			
	Expiry date			
	Quantity contracted			
	Contract price			

VOLUMETRICS

Fig. 12.18

INPUT OUTPUT DESCRIPTION

SYSTEM: SKC- SALES ACCOUNTING	DATE:
AUTHOR: G. CUTTS	PAGE: 3 of 9

NAME: CUSTOMER ORDER/ORDER ACKNOWLEDGEMENT

NARRATIVE: *The order document received from a customer and the context of the order acknowledgement.*

Key	Data item	Format	Len.	Comment

	Customer number
	SKC order number
	  Name
	  Address
	(Order number)
	Date of order
	(Delivery address)
	Dish reference
	Dish description
	Quantity ordered
	(Contract price)
				SKC order number
				Allocated
				consecutively within
				customer to
				make the order
				unique

	VOLUMETRICS		

Fig. 12.19

INPUT OUTPUT DESCRIPTION

SYSTEM: *SKC-SALES ACCOUNTING*	DATE:
AUTHOR: *G. CUTTS*	PAGE: *4* of *9*

NAME: *DELIVERY & DESPATCH NOTE*

NARRATIVE: *A deliverey of one dish to a customer*

Key	Data item	Format	Len.	Comment
	Customer number 　　Name 　　Address (Delivery address) Dish reference Dish description Quantity [to be delivered / delivered] SKC order number Contract delivery			 Yes/No

VOLUMETRICS

Fig. 12.20

INPUT OUTPUT DESCRIPTION

SYSTEM: SKC - SALES & ACCOUNTING	DATE:
AUTHOR: G. CUTTS	PAGE: 5 of 9

NAME: DISH STOCK	

NARRATIVE: A list of the quantities of all dishes put into stock.

Key	Data item	Format	Len.	Comment
	Dish reference			
	Quantity produced			

	VOLUMETRICS		

Fig. 12.21

INPUT OUTPUT DESCRIPTION

SYSTEM: SKC - SALES & ACCOUNTING	DATE:
AUTHOR: G. CUTTS	PAGE: 6 of 9

NAME: PAYMENT

NARRATIVE: A payment received from a customer. Payments are recorded against a specific invoice consecutively.

Key	Data item	Format	Len.	Comment
	Customer number			
	Customer name			
	Invoice number			
	Payment number			Payment number allocated by SKC within the invoice number
	Payment date			
	Payment value			

	VOLUMETRICS	

Fig. 12.22

INPUT OUTPUT DESCRIPTION

SYSTEM: *SKC-SALES & ACCOUNTING*	DATE:
AUTHOR: *G. CUTTS*	PAGE: *7* of *9*

NAME: *SALES INVOICE*

NARRATIVE: *An invoice comprising many invoice lines where each invoice line represents a deliverey of a dish*

Key	Data item	Format	Len.	Comment
	Customer number			
	Customer name			
	Customer address			
	Invoice number			
	Invoice date			
	SKC order number			
	Dish reference			
	Dish description			
	Standard price			
	Contract price			
	Quantity delivered			
	Invoice line value			
	Total invoice value			

VOLUMETRICS

Fig. 12.23

INPUT OUTPUT DESCRIPTION

| SYSTEM: SKC - SALES & ACCOUNTING | DATE: |
| AUTHOR: G. CUTTS | PAGE: 8 of 9 |

NAME: SALES REPORT/MANAGEMENT REPORT

NARRATIVE: *For each customer produce the following report.*

Key	Data item	Format	Len.	Comment
	Customer number			
	Name			From the
	Dish reference			order-line entity
	Dish description			
	SKC order number			
	Qty. ordered			
	Order stopped			Yes/No
	(Qty. delivered)			
	(Invoice line value)			
	(Contract number)			
	(Contract price)			

VOLUMETRICS

Fig. 12.24

```
┌─────────────────────────────────────────────────────────┐
│ ┌─────────────────────────────────────────────────────┐ │
│ │ INPUT OUTPUT DESCRIPTION                              │ │
│ └─────────────────────────────────────────────────────┘ │
```

SYSTEM: *SKC - SALES ACCOUNTING*	DATE:
AUTHOR: *G. CUTTS*	PAGE: *9* of *9*

NAME: *SALES STATEMENT*	

NARRATIVE: *A statement of the current invoices, payments and credit notes.*

Key	Data item	Format	Len.	Comment
	Customer number			
	Name			
	Address			
	Invoice number			
	Date			
	Total invoice value			
	(Credit note number			
	Credit note date			
	Credit value)			
	(Payment number			
	Payment date			
	Payment value)			

	VOLUMETRICS		

Fig. 12.25

12.6 SKC: DATA DICTIONARY, TASK 2.8

Figure 12.26 shows an initial entry made in the data dictionary for the attribute 'customer name' from the customer entity. At this time it has been entered into the data dictionary 'undefined'. Figure 12.27 shows the final entry for 'customer name' and Figure 12.28 shows an entry for 'contract price' from the contract entity.

12.7 SKC: ON-LINE DIALOGUE SPECIFICATION, TASK 2.9

Figure 12.29 shows the on-line dialogue specification. It has been designed directly from the data flow diagrams. A main menu allows access to on-line functions for sales and accounting and to reporting functions. In each case a second level menu is displayed. The second level provides access to the individual functions taken from the level 2 data flow diagrams.

12.8 SKC: FUNCTION DESCRIPTIONS, TASK 2.10

Figures 12.30 to 12.33 documents each primitive DFD function.

DATA DICTIONARY

SYSTEM: SKC	DATE: 12/01/90
AUTHOR: G. CUTTS	PAGE: I of: I

DATA ITEM: CUSTOMER   NAME

SHORT NAME: UNDEFINED

Type:

Length:

Format:

Characteristics

SIGNED:	YES/NO
ROUNDED:	YES/NO
PACKED:	YES/NO
JUSTIFIED:	YES/NO

Range:

Description:

Fig. 12.26

DATA DICTIONARY

SYSTEM: SKC	DATE: 12/01/90
AUTHOR: G. CUTTS	PAGE: 1 of: 1

DATA ITEM: CUSTOMER NAME

SHORT NAME: CUST- NAME

Type: ALPHA - NUMERIC

Length: 20 BYTES

Format: —

Characteristics

SIGNED:	YES/NO
ROUNDED:	YES/NO
PACKED:	YES/NO
JUSTIFIED:	YES/NO LEFT

Range:

Description: THE CUSTOMER NAME USED ON ALL
OUTPUT.

Fig. 12.27

DATA DICTIONARY

SYSTEM: SKC	DATE: 12/01/90
AUTHOR: G.CUTTS	PAGE: 1 of: 1

DATA ITEM: CONTRACT PRICE

SHORT NAME: CONT - PRICE

Type: NUMERIC

Length: 5 BYTES

Format: 999.99

Characteristics

SIGNED·	YES/NO
ROUNDED:	YES/NO
PACKED:	YES/NO
JUSTIFIED:	YES/NO

Range: 000.00 ——— 199.99

Description: THE SPECIAL PRICE FOR A DISH NEGOTIATED ON A CONTRACT BY A CUSTOMER.

Fig. 12.28

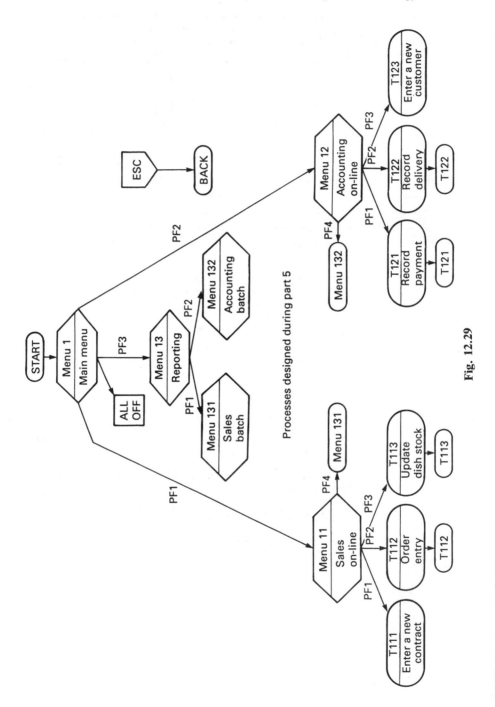

Processes designed during part 5

Fig. 12.29

LOGICAL FUNCTION DESCRIPTIONS

SYSTEM: *SKC-SALES ACCOUNTING*	DATE:
AUTHOR: *G. CUTTS*	PAGE: *1* of *5*

DFD Ref.	Name	Description
1.1	Process contract	Insert contract details into the system.
1.2	Record order & qty. ordered against contract	Insert sales order details into the system. If any dish ordered contains a contract price, record the quantity ordered for that dish on the appropriate contract.
1.3	Check credit	For all orders inserted by function 1.2, check that the customer is not stop listed. Modify the order to reflect the stop list status.
1.4	Check stock availability	For each dish on the order which has a pending status or outstanding status, check if there is sufficient stock for the order. If stock is sufficient, then mark the order for despatch (i.e. change its status) else mark the status as pending. All pending orders must be processed before outstanding orders. Decrement the stock for dishes despatches.

Fig. 12.30

LOGICAL FUNCTION DESCRIPTIONS

SYSTEM: SKC-SALES ACCOUNTING	DATE:
AUTHOR: G. CUTTS	PAGE: 2 of 5

DFD Ref.	Name	Description
1.5	Produce order acknowledgement	For all orders not previously acknowledged, generate an order acknowledgement.
1.6	Produce despatch note	For all order lines with status ready to despatch, produce a despatch note. Change the status to indicate that a despatch note has been produced.
1.7	Update dish stock	Increment the dish quantity in stock for all dishes recorded on the dish production list.

Fig. 12.31

LOGICAL FUNCTION DESCRIPTIONS

SYSTEM: SKC - SALES ACCOUNTING	DATE:
AUTHOR: G. CUTTS	PAGE: 3 of 5

DFD Ref.	Name	Description
2.1	Record payment	For each payment received, examine the latest sales statement, allocate the payment to an invoice and insert the payment into the system.
2.2	Record new customer details	Insert newly authorised customer details into the system.
2.3	Produce credit notes	For each credit authorisation insert the credit into the system and generate a credit note for the customer. Credits must be allocated to invoices.
2.4	Select customers for stop list	Examine all customer accounts. Where a customer has exceeded the credit limit or credit time and has not obtained management clearance, then the customer status must be changed to stop listed.

Credit limit exceeded	Y	—	—	N
Credit time exceeded	—	Y	—	N
Management exception	N	N	Y	N
Stop list	X	X		
OK			X	X

Fig. 12.32

LOGICAL FUNCTION DESCRIPTIONS

SYSTEM: *SKC-SALES ACCOUNTING*	DATE:
AUTHOR: *G. CUTTS*	PAGE: 4 of 5

DFD Ref.	Name	Description
2.5	Authorise credit	For all returned dishes, authorise credit according to management guidelines.
2.6	Create invoice line	For all deliveries, insert an invoice line into the system. For deliveries against a contract, use the contract price, for all other deliveries use the standard price to generate the invoice line value.
2.7	Produce sales statement	Generate a sales statement for each customer showing all outstanding invoices, credit notes and payments.
2.8	Produce invoice	For all invoice lines inserted into the system, with a status of not printed, generate sales invoice showing all the invoice lines. Insert the invoice head into the system.

LOGICAL FUNCTION DESCRIPTIONS

SYSTEM: *SKC-SALES ACCOUNTING*	DATE:
AUTHOR: *G. CUTTS*	PAGE: 5 of 5

DFD Ref.	Name	Description
3	Produce management report	For all customers, generate a report showing for each dish ordered by the customer the quantity ordered, and the quantity invoiced. Also show if the dish was a standard or contract order and the contract price. Indicate all orders stopped because of credit violations.

Fig. 12.33

12.9 SKC: ENTITY FUNCTION MATRIX, TASK 2.11

The entity function matrix shown in Figure 12.34 was constructed directly from the required DFDs and the required entity model. By reading across the rows, it can be easily seen that the following functions are required:

(1) A function to insert and modify the entity dish.

(2) No entities are deleted or archived from the database.

Additional functions and modifications to existing functions are required as follows:

- To allow the dish maintenance function to delete obsolete entity occurrences.
- To allow function 2.7, produce sales statements, to delete all invoice heads, invoice lines, credit notes and payments when the total credit note value plus payment value equals the invoice value and when all entities have been printed on a sales statement.
- To create a new function to delete order head and order line entities when all order lines for the order have been invoiced. This requires a modification to function 2.8, produce invoice.
- A monthly review of customers and contracts should provide for the archiving of inactive customer entities and completed contracts.

The consequence of the above is three new functions and two modified functions.

New functions
Function 4.1 Maintain dish
Function 4.2 Archive order head and order line
Function 4.3 Archive customer and contract

Revised functions
Function 2.7 Allow deletion of invoice head, invoice line, credit note and payment
Function 2.8 Modify order lines to show that they have been invoiced.

The new and revised functions are documented and shown in Figure 12.35. A revised entity function matrix is shown in Figures 12.36(a) and (b). It is also necessary to update the required DFDs with the new and revised functions. Figures 12.37(a) and (b) show the revised required DFDs. There are no changes to pages 2 and 3 of the DFDs.

ENTITY/FUNCTION MATRIX

SYSTEM: SKC - SALES ACCOUNTING	DATE:
AUTHOR: G. CUTTS	PAGE: 1 of 1

Entity name \ Function name	1.1 Process contract	1.2 Record order & qty. against contract	1.3 Check credit	1.4 Check stock availability	1.5 Produce order acknowledgement	1.6 Produce despatch note	1.7 Update dish stock	2.1 Record payment	2.2 Record new customer details	2.3 Produce credit note	2.4 Select customers for stop list	2.5 Authorise credit	2.6 Create invoice line	2.7 Produce sales statement	2.8 Produce invoice	3 Produce management report
Contract	I	M											R			R
Credit note									I	R				R		
Customer		R		R	R				I	R	M			R	R	R
Dish				M			M						R			
Invoice head								R			R	R			M	I
Invoice line														I	M	R
Order head	I	M		M	R											R
Order line		I		M	R	M										R
Payment								I				R		R		

Fig. 12.34

LOGICAL FUNCTION DESCRIPTIONS

SYSTEM: *SKC-SALES ACCOUNTING*	DATE:
AUTHOR: *G. CUTTS*	PAGE: / of /

DFD Ref.	Name	Description
2.7	Produce sales statement (additional)	Printing must continue until invoice value $\leq$ credit value + payment value then delete invoice head, invoice line, credit & payment & print the final values.
2.8	Produce invoice (additional)	Modify the order line to indicate an invoice has been produced.
4.1	Maintain dish	This function inserts, modifies or deletes all attributes associated with the entity dish.
4.2	Archive order head & order line	Archive (delete from the system) all order & order line entities when all order lines have been invoiced.
4.3	Achieve customer & contract	Archive (delete from the system) all customer and contract entities where there has been no activity for six months and all contracts have expired.

Fig. 12.35

ENTITY/FUNCTION MATRIX

SYSTEM: SKC - SALES ACCOUNTING	DATE:
AUTHOR: G. CUTTS	PAGE: 1 of 2

Entity name / Function name	1.1 Process contract	1.2 Record order & qty. against contract	1.3 Check credit	1.4 Check stock availability	1.5 Produce order acknowledgement	1.6 Produce despatch note	1.7 Update dish stock	2.1 Record payment	2.2 Record new customer details	2.3 Produce credit note	2.4 Select customer for stop list	2.5 Authorise credit	2.6 Create invoice line	2.7 Produce sales statement	2.8 Produce invoice	3 Produce management report
Contract	I	M											R			R
Credit note										I	R			R/A		
Customer		R		R	R				I	R	M			R	R	R
Dish				M			M						R			
Invoice head								R			R	R		M/A	I	
Invoice line													I	A	M	R
Order head		I	M		M	R										R
Order line		I		M	R	M									M	R
Payment								I			R			R/A		

Fig. 12.36(a)

ENTITY/FUNCTION MATRIX

SYSTEM: SKC - SALES ACCOUNTING	DATE:
AUTHOR: G. CUTTS	PAGE: 2 of 2

Entity name \ Function name	4.1 Maintain dish	4.2 Archive order & order line	4.3 Archive customer & contract													
Contract			A													
Credit note																
Customer			A													
Dish	I/M/D															
Invoice head																
Invoice line																
Order head		A														
Order line		A														
Payment																

Fig. 12.36(b)

DATA FLOW DIAGRAM

SYSTEM: *SKC –SALES &ACCOUNTING*	DATE:
AUTHOR: *G. CUTTS*	PAGE: / of 4

LEVEL: 1	~~CURRENT~~/REQ.	PHYS./~~LOGICAL~~

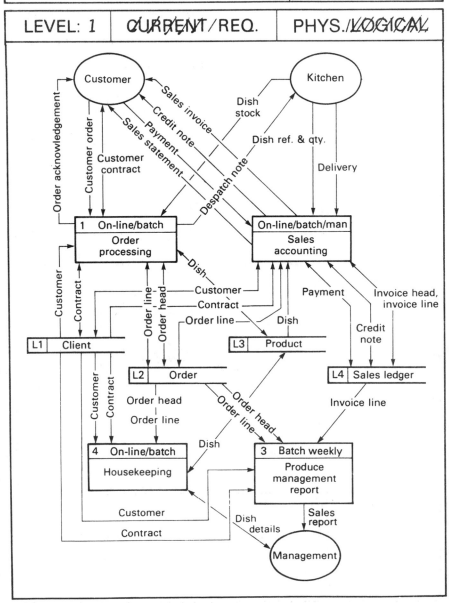

Fig. 12.37(a)

DATA FLOW DIAGRAM

SYSTEM: *SKC SALES & ACCOUNTING*	DATE:
AUTHOR: *G. CUTTS*	PAGE: 4 of 4

LEVEL: 2	~~CURRENT~~/REQ.	PHYS./~~LOGICAL~~

TITLE: HOUSEKEEPING

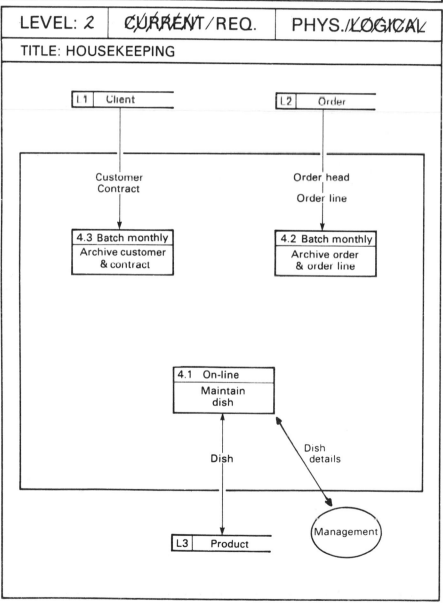

Fig. 12.37(b)

12.10 SKC: ENTITY LIFE HISTORIES, TASK 2.12

Figures 12.38 to 12.49 document the entity life histories for the nine entities. Three entity life histories have been enhanced to show possible abnormal lives. The order head life can deadlock at the function check credit, by continual failure. Two new functions are required. Function 1.9 rejects and deletes the order and function 1.10 accepts the order without passing the credit check. An order may also be cancelled by the customer. Cancellation may be processed from any status. Order lines must not be processed until the order has been cleared for credit.

If an order is rejected, function 1.9, then the order lines must be deleted. Similarly, if an order is cancelled, then, again, the order lines must be deleted. Cancel order is only acceptable before a despatch note is produced. Order line status 1 and 2 are, therefore, connected to function 1.8, cancel order. Function 1.8 cannot execute if any order line is in status 3 or 4.

If an order is cancelled after the allocation of stock, function 1.4, then this must be notionally returned to stock. Function 1.8, cancel order, must modify the dish entity.

An entity function matrix is shown in Figure 12.47 for functions 1.8, cancel order; 1.9, reject order; 1.10, accept order; and 1.3, check credit.

The logical function descriptions for functions 1.8, 1.9 and 1.10 are shown in Figure 12.48.

The revised level 2 required DFD incorporating functions 1.8, 1.9 and 1.10 is shown in Figure 12.49.

ENTITY LIFE HISTORY

SYSTEM: *SKC - SALES ACCOUNTING* | DATE:

AUTHOR: *G. CUTTS* | PAGE: *1* of *9*

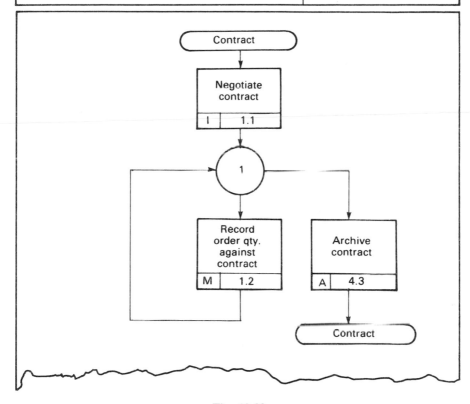

Fig. 12.38

ENTITY LIFE HISTORY

SYSTEM: SKC- SALES & ACCOUNTING	DATE:
AUTHOR: G. CUTTS	PAGE: 2 of 9

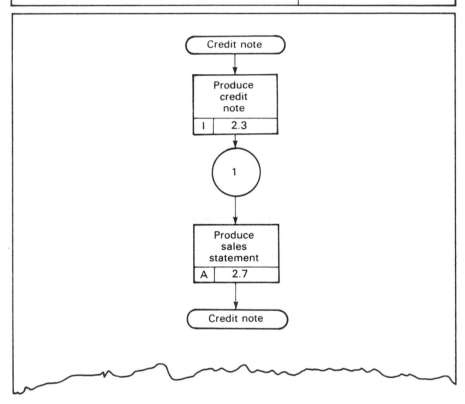

Fig. 12.39

ENTITY LIFE HISTORY

SYSTEM: SKC - SALES AND ACCOUNTING | DATE:

AUTHOR: G. CUTTS | PAGE: 3 of 9

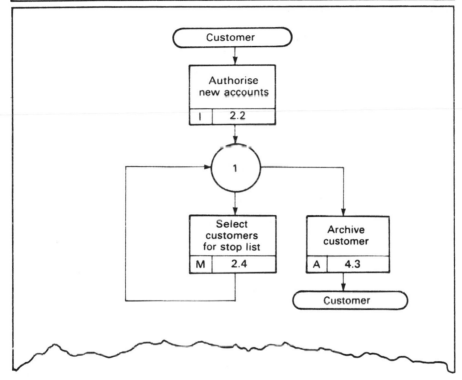

Fig. 12.40

ENTITY LIFE HISTORY

SYSTEM:SKC-SALES AND ACCOUNTING	DATE:
AUTHOR: G. CUTTS	PAGE: 4 of 9

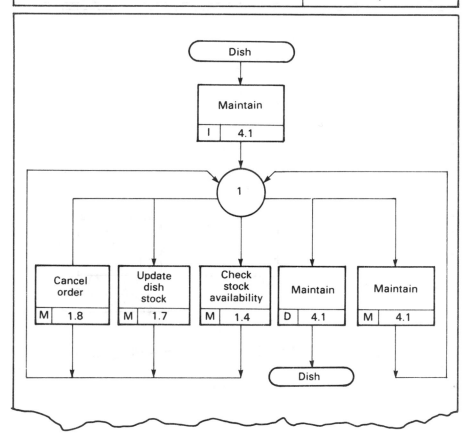

Fig. 12.41

ENTITY LIFE HISTORY

SYSTEM: SKC-SALES AND ACCOUNTING | DATE:

AUTHOR: G. CUTTS | PAGE: 5 of 9

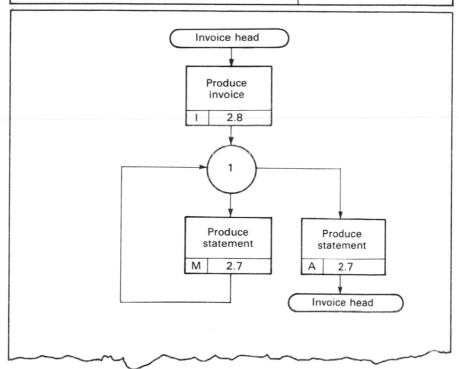

Fig. 12.42

ENTITY LIFE HISTORY

SYSTEM: SKC-SALES AND ACCOUNTING	DATE:
AUTHOR: G. CUTTS	PAGE: 6 of 9

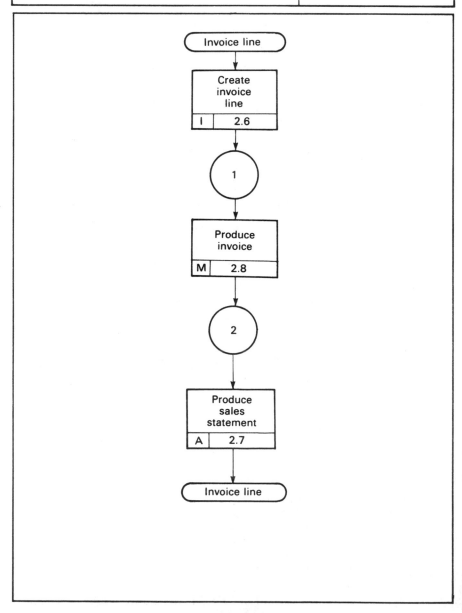

Fig. 12.43

ENTITY LIFE HISTORY

SYSTEM: *SKC- SALES AND ACCOUNTING*	DATE:
AUTHOR: *G. CUTTS*	PAGE: 7 of 9

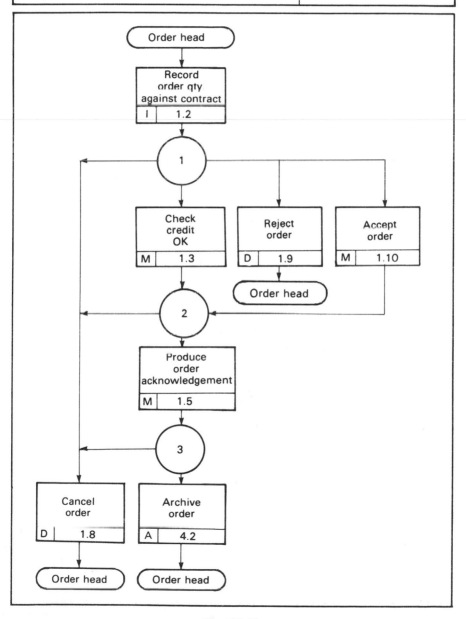

Fig. 12.44

ENTITY LIFE HISTORY

SYSTEM: SKC-SALES & ACCOUNTING	DATE:
AUTHOR: G. CUTTS	PAGE: 8 of: 9

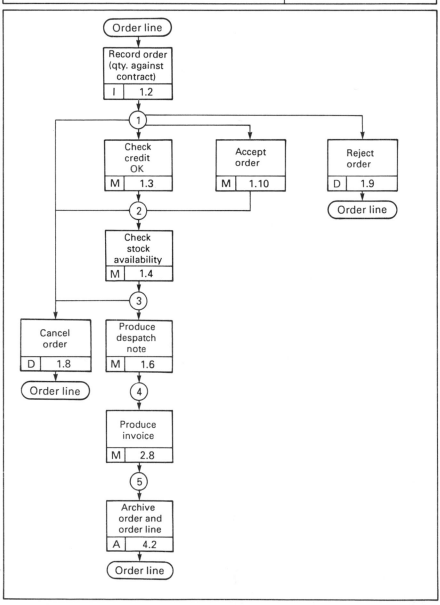

Fig. 12.45

ENTITY LIFE HISTORY

SYSTEM: SKC-SALES & ACCOUNTING

DATE:

AUTHOR: G. CUTTS

PAGE: 9 of 9

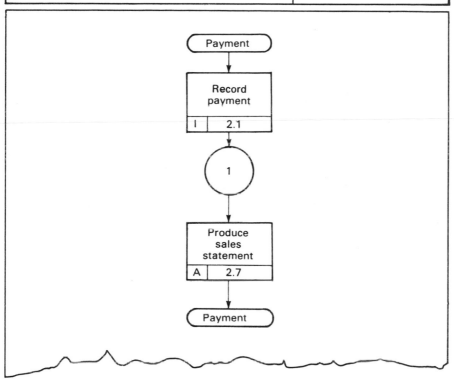

Fig. 12.46

ENTITY/FUNCTION MATRIX

SYSTEM: SKC - SALES AND ACCOUNTING												DATE:		
AUTHOR: G. CUTTS												PAGE: 1 of 1		

Event name / Entity name	1.8 Cancel order	1.9 Reject Order	1.10 Accept order		1.3 Check credit								
Contract													
Credit note													
Customer					R								
Dish	M												
Invoice head													
Invoice line													
Order head	D	D	M		M								
Order line	D	D	M		M								
Payment													

Fig. 12.47

LOGICAL FUNCTION DESCRIPTIONS

SYSTEM: SKC-SALES AND ACCOUNTING	DATE:
AUTHOR: G. CUTTS	PAGE: 1 of 1

DFD Ref.	Name	Description
1.8	Cancel order	Delete the order head and order lines from the system. If stock has been allocated to the order line, increment the dish stock with the quanity allocated.
1.9	Reject order	Delete the order head and order lines.
1.10	Accept order	Modify the order head and order line to indicate that processing may continue
1.3	Check credit	For all orders check if the customer is stop listed. If not stop listed modify the order head and order line to allow continued processing.

Fig. 12.48

DATA FLOW DIAGRAM

SYSTEM: *SKC SALES & ACCOUNTING*   DATE:

AUTHOR: *G. CUTTS*   PAGE: 2 of 4

LEVEL: 2 | C̶U̶R̶R̶E̶N̶T̶/REQ. | PHYS./L̶O̶G̶I̶C̶A̶L̶

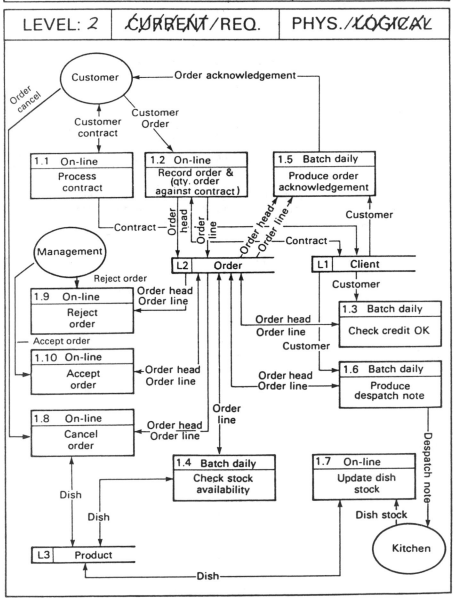

Fig. 12.49

Chapter 13
SKC Part 3: Logical Data Design

13.1 SKC: SELECTED DATA STRUCTURES, TASK 3.1

All the input and output to the required computer functions was selected for normalisation.

13.2 SKC: DATA STRUCTURES IN 3NF, TASK 3.2

The selected data structures have been normalised to produce third normal from data structures. These data structures are documented in Figures 13.1 to 13.9.

13.3 SKC: ENTITY DESCRIPTIONS, TASK 3.3

The third normal form data structures have been merged to produce nine entity descriptions which are shown in Figures 13.10 to 13.18.

NORMALISATION

SYSTEM: SKC SALES & ACCOUNTING	DATE:
AUTHOR: G. CUTTS	PAGE: 1 of 18

DATA STRUCTURE: *CREDIT NOTE*

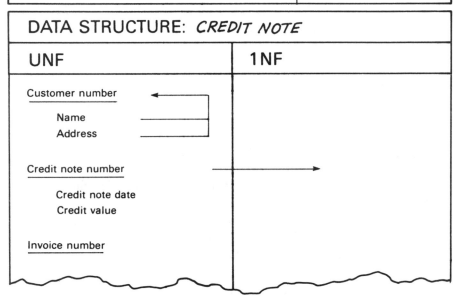

UNF	1NF
Customer number	
Name	
Address	
Credit note number	
Credit note date	
Credit value	
Invoice number	

Fig. 13.1(a)

NORMALISATION

SYSTEM: Skc SALES & ACCOUNTING | DATE:

AUTHOR: G. CUTTS | PAGE: 2 of 18

DATA STRUCTURE: CREDIT NOTE

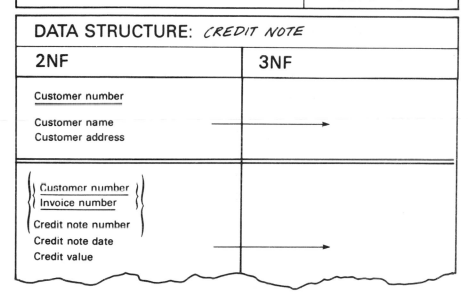

Fig. 13.1(b)

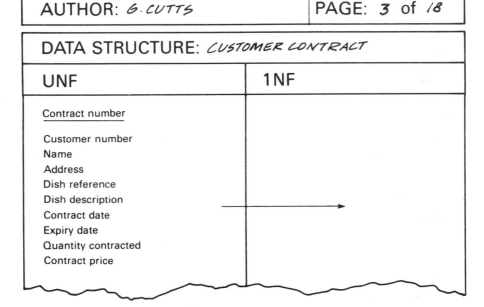

NORMALISATION

SYSTEM: *SALES ACCOUNTING*	DATE:
AUTHOR: *G. CUTTS*	PAGE: *3* of *18*

DATA STRUCTURE: *CUSTOMER CONTRACT*

UNF	1NF
<u>Contract number</u> Customer number Name Address Dish reference Dish description Contract date Expiry date Quantity contracted Contract price	

Fig. 13.2(a)

NORMALISATION

SYSTEM: *SALES ACCOUNTING*	DATE:
AUTHOR: *G. CUTTS*	PAGE: *4* of *18*

DATA STRUCTURE: *CUSTOMER CONTRACT*

2NF	3NF
2NF is identical to 1NF since there is only a simple key <u>Contract number</u> Customer number Name Address Dish reference Dish description Contract date Expiry date Quantity contracted Contract price	<u>Customer number</u> Name Address --- <u>Dish reference</u> Dish description --- <u>Contract number</u> Customer number* Dish reference* Contract date Expiry date Quantity contracted Contract price

Fig. 13.2(b)

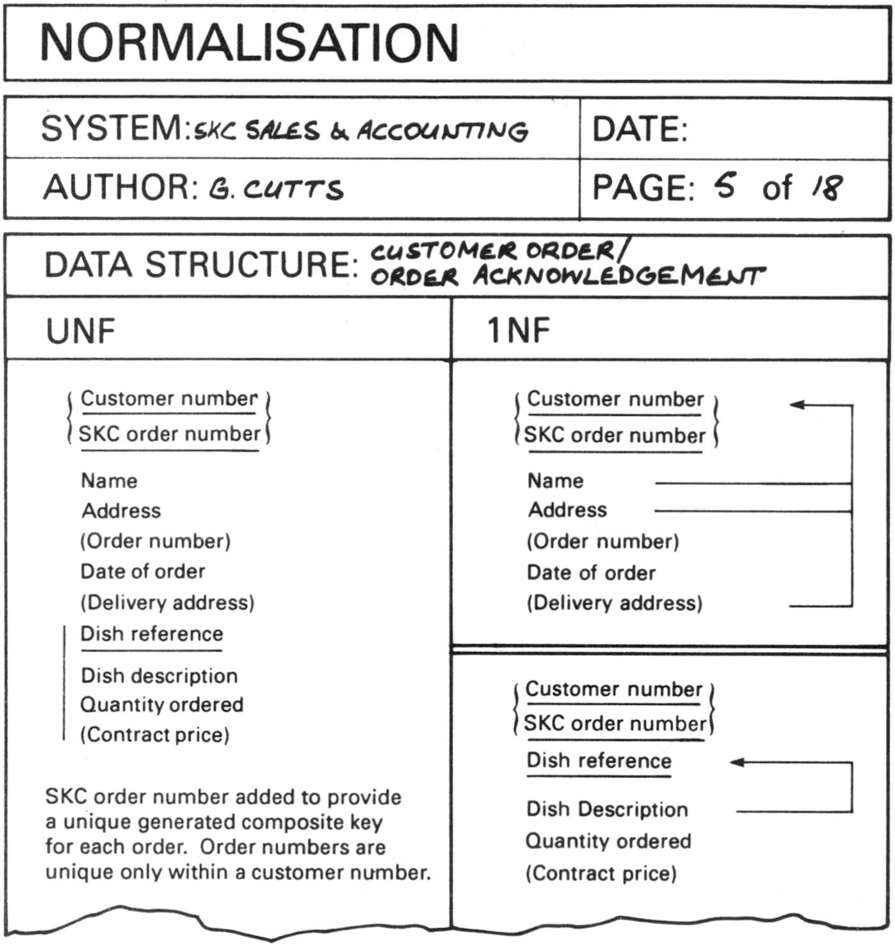

NORMALISATION

SYSTEM: *SKC SALES & ACCOUNTING*   DATE:

AUTHOR: *G. CUTTS*   PAGE: *5* of *18*

DATA STRUCTURE: *CUSTOMER ORDER/ ORDER ACKNOWLEDGEMENT*

UNF	1NF
{ Customer number } { SKC order number } Name Address (Order number) Date of order (Delivery address) Dish reference Dish description Quantity ordered (Contract price) SKC order number added to provide a unique generated composite key for each order. Order numbers are unique only within a customer number.	{ Customer number } { SKC order number } Name Address (Order number) Date of order (Delivery address) --- { Customer number } { SKC order number } Dish reference Dish Description Quantity ordered (Contract price)

Fig. 13.3(a)

NORMALISATION

SYSTEM: *SALES ACCOUNTING*	DATE:
AUTHOR: *G. CUTTS*	PAGE: 6 of *18*

DATA STRUCTURE: *CUSTOMER ORDER* / *ORDER ACKNOWLEDGMENT*

2NF	3NF
<u>Customer number</u> Name Address (Delivery address)	
{ <u>Customer number</u> } { SKC order number } (Order number) Date of order	
<u>Dish reference</u> Dish description	
{ <u>Customer number</u> } { SKC order number } <u>Dish reference</u> Quantity ordered (Contract price)	{ <u>Customer number</u> } { SKC order number } <u>Dish reference</u> Quantity ordered (Contract number)*
	The contract price is replaced by its implied key contract number, which becomes a foreign key.

Fig. 13.3(b)

NORMALISATION

SYSTEM: *SKC SALES & ACCOUNTING*   DATE:

AUTHOR: *G. CUTTS*   PAGE: 7 of *18*

DATA STRUCTURE: *DELIVERY/DESPATCH NOTE*

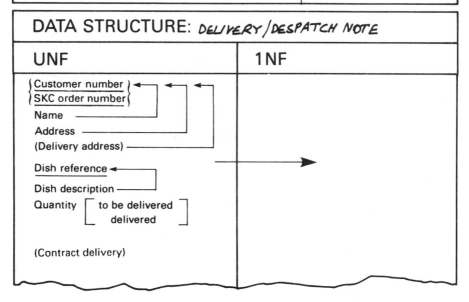

UNF	1NF

Fig. 13.4(a)

NORMALISATION

SYSTEM: *SKC SALES & ACCOUNTING*	DATE:
AUTHOR: *G. CUTTS*	PAGE: *8* of *18*

DATA STRUCTURE: *DELIVERY/DESPATCH*

2NF	3NF
<u>Customer number</u> Name Address (Delivery address) ⟶	
<u>Dish reference</u> Dish description ⟶	
{ <u>Customer number</u> } { <u>SKC order number</u> } Dish reference ⌐ Quantity to be delivered ⌐ ⌐ Quantity delivered ⌐ (Contract delivery)	{ <u>Customer number</u> } { <u>SKC order number</u> } <u>Dish reference</u> ⌐ Quantity to be delivered ⌐ ⌐ Quantity delivered ⌐ (Contract number)*

Fig. 13.4(b)

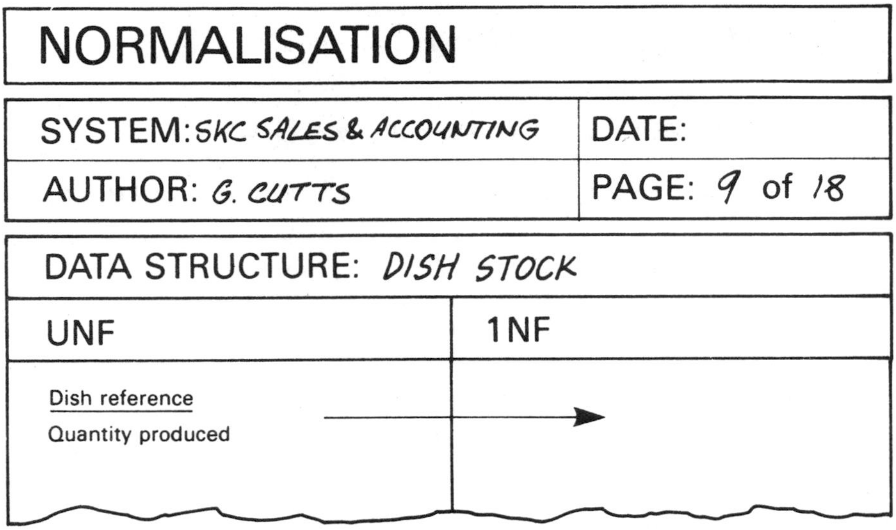

NORMALISATION

SYSTEM: *SKC SALES & ACCOUNTING*	DATE:
AUTHOR: *G. CUTTS*	PAGE: *9* of *18*

DATA STRUCTURE: *DISH STOCK*

UNF	1NF
Dish reference Quantity produced	

Fig. 13.5(a)

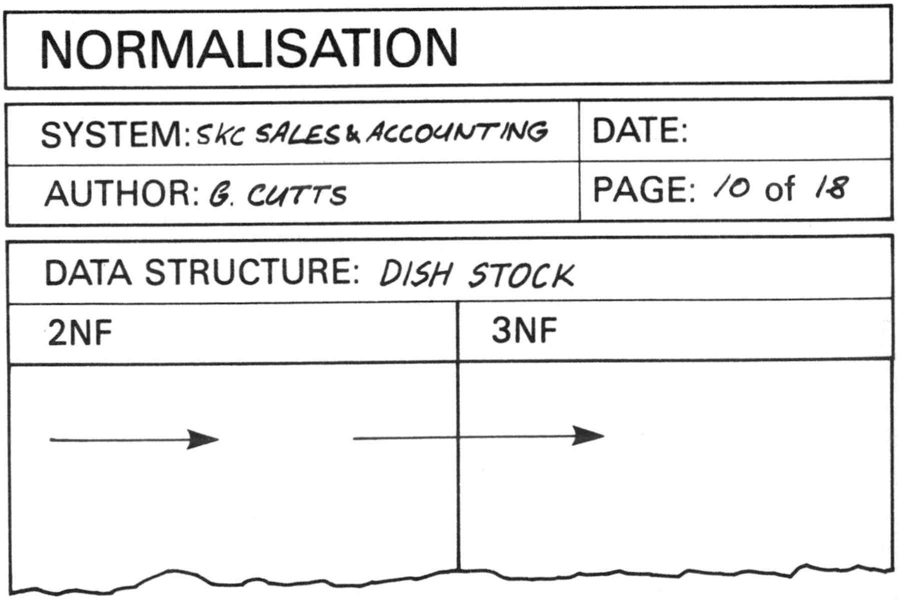

NORMALISATION

SYSTEM: *SKC SALES & ACCOUNTING*	DATE:
AUTHOR: *G. CUTTS*	PAGE: *10* of *18*

DATA STRUCTURE: *DISH STOCK*

2NF	3NF

Fig. 13.5(b)

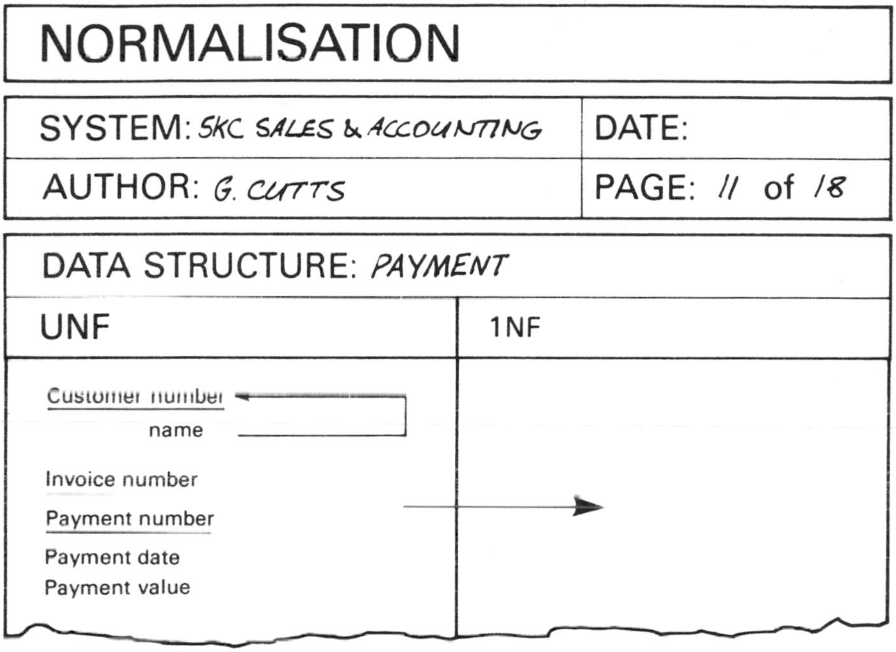

NORMALISATION

SYSTEM: *SKC SALES & ACCOUNTING*	DATE:
AUTHOR: *G. CUTTS*	PAGE: *11* of *18*

DATA STRUCTURE: *PAYMENT*

UNF	1NF
Customer number name Invoice number Payment number Payment date Payment value	

Fig. 13.6(a)

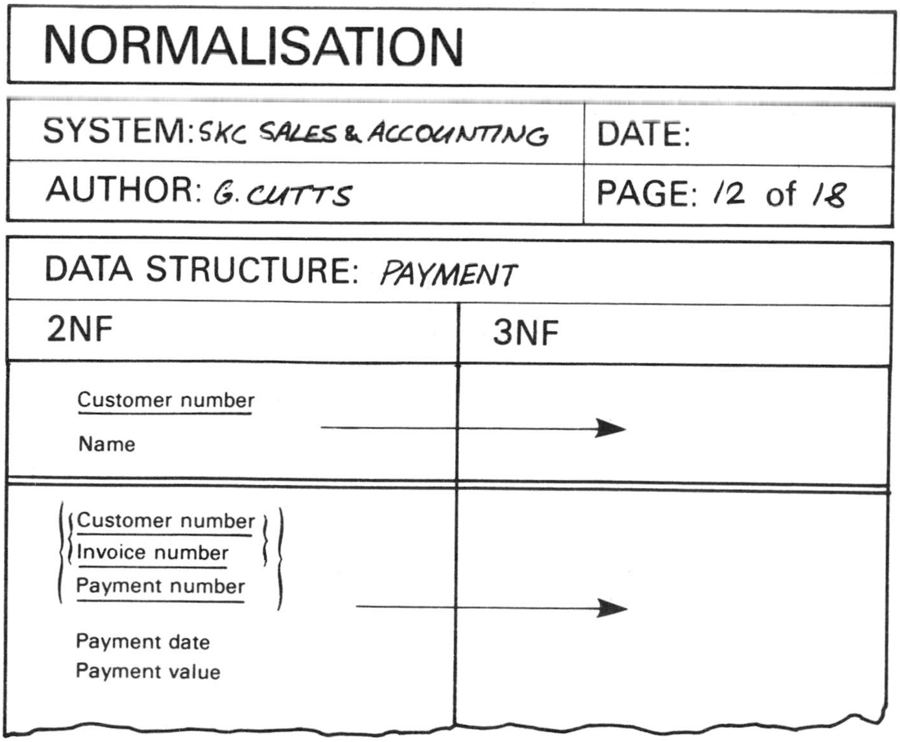

NORMALISATION

SYSTEM: *SKC SALES & ACCOUNTING*	DATE:
AUTHOR: *G. CUTTS*	PAGE: *12* of *18*

DATA STRUCTURE: *PAYMENT*

2NF	3NF
Customer number Name	
{Customer number} {Invoice number} Payment number Payment date Payment value	

Fig. 13.6(b)

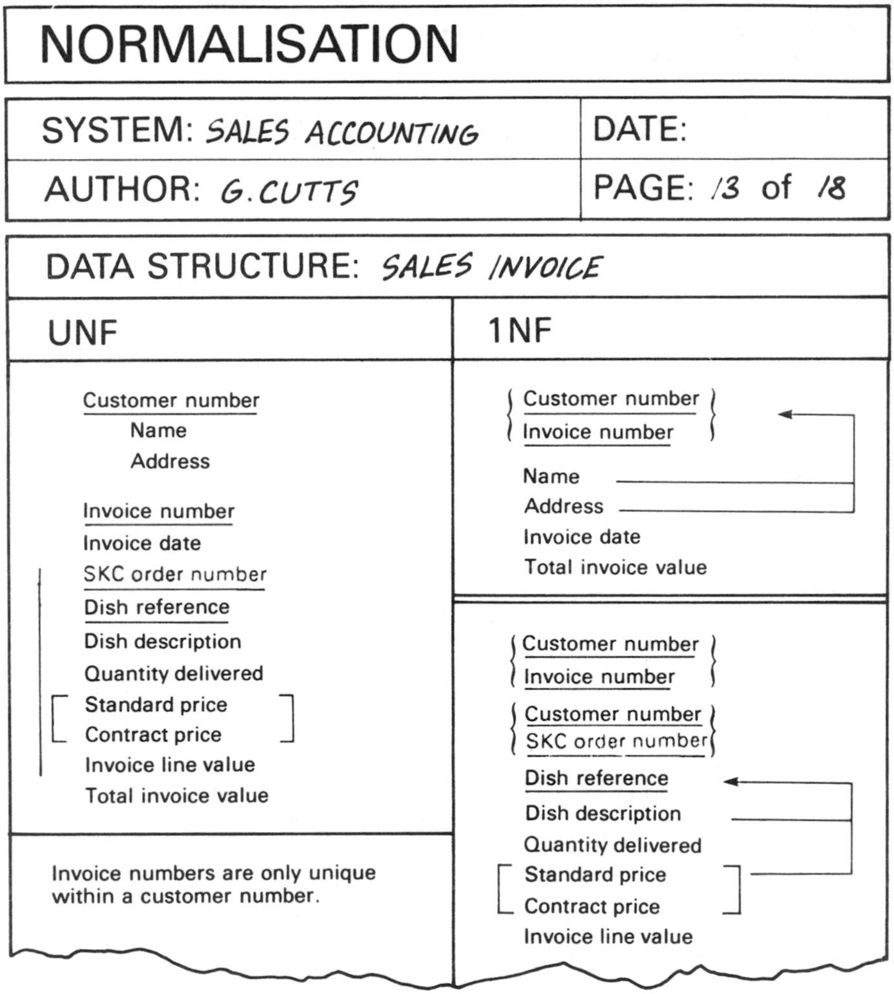

NORMALISATION

SYSTEM: SALES ACCOUNTING	DATE:
AUTHOR: G.CUTTS	PAGE: /3 of /8

DATA STRUCTURE: SALES INVOICE

UNF	1NF

Customer number
  Name
  Address
 	
Invoice number
Invoice date
SKC order number
Dish reference
Dish description
Quantity delivered
Standard price
Contract price
Invoice line value
Total invoice value
	{ Customer number }
	(Invoice number } ←
	
	Name
	Address
	Invoice date
	Total invoice value

Invoice numbers are only unique within a customer number.	{ Customer number } (Invoice number } { Customer number } (SKC order number(Dish reference ← Dish description Quantity delivered Standard price Contract price Invoice line value

Fig. 13.7(a)

NORMALISATION

SYSTEM: *SALES ACCOUNTING*	DATE:
AUTHOR: *G. CUTTS*	PAGE: *14* of *18*

DATA STRUCTURE: *SALES INVOICE*

2NF	3NF
Customer number Name Address ⟶	⟶
{ Customer number Invoice number } Invoice date Total invoice value ⟶	⟶
Dish reference Dish description ⟶ Standard price	⟶

{ Customer number
  Invoice number }
 	
{ Customer number
  SKC order number }
 	
Dish reference
Quantity delivered
Invoice line value
(Contract price)
	{ Customer number
	  Invoice number }
	
	Dish reference
	
	{ Customer number
	  SKC order number }
	
	Quantity delivered
	Invoice line value
	(Contract number)*
	
	Contract price replaced
	by Contract number.

Fig. 13.7(b)

NORMALISATION

SYSTEM: SKC SALES & ACCOUNTING	DATE:
AUTHOR: G. CUTTS	PAGE: 15 of 18

DATA STRUCTURE: SALES REPORT

UNF	1NF
<u>Customer Number</u> name	<u>Customer number</u> Name
<u>Dish reference</u> Dish description SKC order number Quantity ordered Order stopped (Qty. delivered) (Invoice line value) (Contract number) (Contract price)	Customer number SKC order number Dish reference Dish description Qty. ordered Order stopped (Qty. delivered) (Invoice line value) (Contract number) (Contract price)

Fig. 13.8(a)

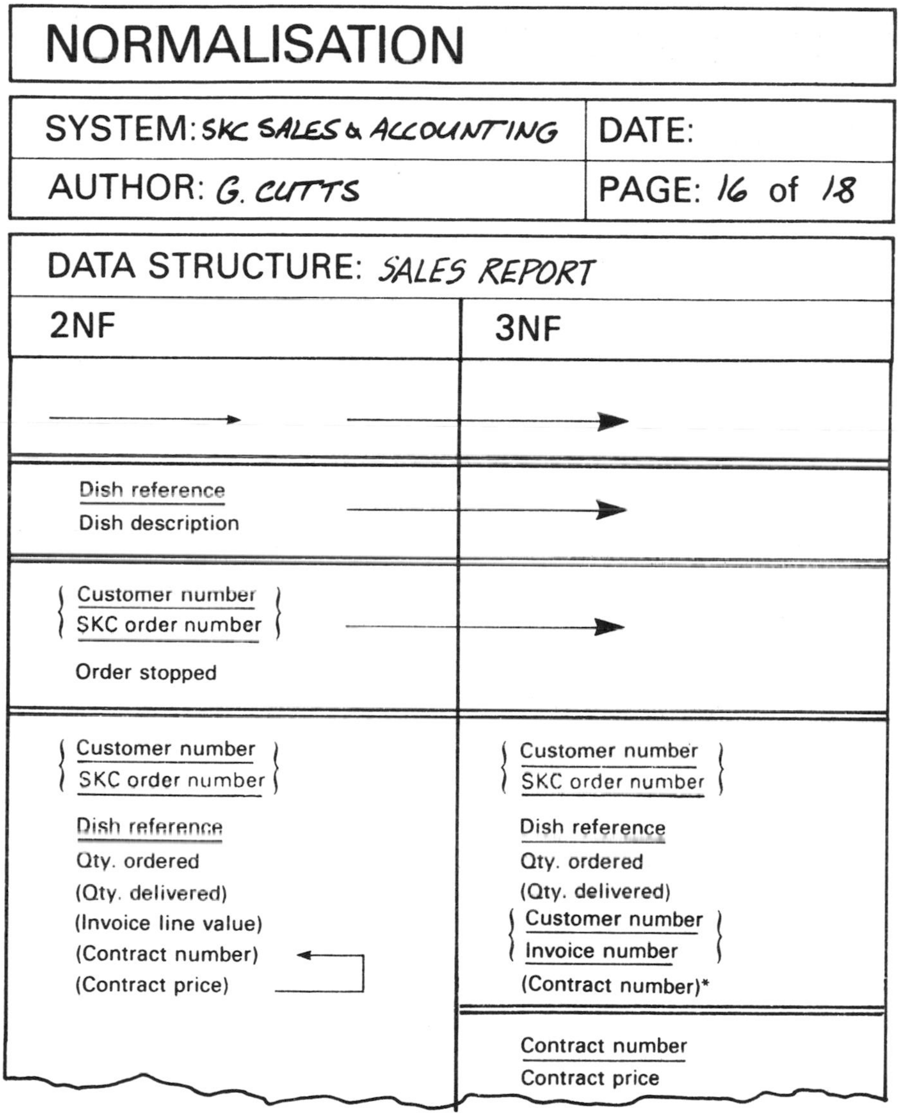

NORMALISATION

SYSTEM: SKC SALES & ACCOUNTING	DATE:
AUTHOR: G. CUTTS	PAGE: 16 of 18

DATA STRUCTURE: SALES REPORT

2NF	3NF
⟶	⟶
<u>Dish reference</u> Dish description	⟶
{ <u>Customer number</u> { <u>SKC order number</u> Order stopped	⟶
{ <u>Customer number</u> { <u>SKC order number</u> <u>Dish reference</u> Qty. ordered (Qty. delivered) (Invoice line value) (Contract number) (Contract price)	{ <u>Customer number</u> { <u>SKC order number</u> <u>Dish reference</u> Qty. ordered (Qty. delivered) { <u>Customer number</u> { <u>Invoice number</u> (Contract number)*
	<u>Contract number</u> Contract price

Fig. 13.8(b)

NORMALISATION

SYSTEM: *SALES ACCOUNTING*	DATE:
AUTHOR: *G. CUTTS*	PAGE: *17* of *18*

DATA STRUCTURE: *SALES STATEMENT*

UNF	1NF

Customer number
  Name
  Address
 	
Invoice number
Invoice date
Total invoice value
 	
(Credit note number
  Credit note date
  Credit value)
(Payment number
  Payment date
  Payment value)
	Customer number
	  Name
	  Address
	
	{ Customer number
	  Invoice number }
	
	Invoice date
	Total invoice number
	
	{{Customer number}
	{Invoice number}
	{ Credit·note number }
	
	Credit note date
	Credit value
	
	{{Customer number}}
	{{Invoice number}}
	{ Payment number }
	
	Payment Date
	Payment Value

Fig. 13.9(a)

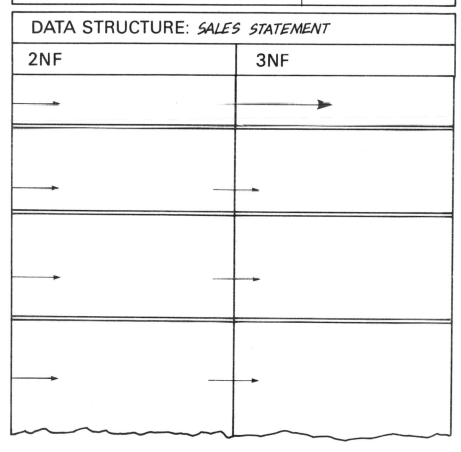

NORMALISATION

SYSTEM: *SALES ACCOUNTING*	DATE:
AUTHOR: *G. CUTTS*	PAGE: *18* of *18*

DATA STRUCTURE: *SALES STATEMENT*

2NF	3NF

Fig. 13.9(b)

ENTITY DESCRIPTION

SYSTEM: SKC SALES & ACCOUNTING	DATE:
AUTHOR: G. CUTTS	PAGE: 1 of 9

NAME: CONTRACT

NARRATIVE:

Key	Data item	Format	Len.	Comment
✓	Contract number			
	Customer number*			
	Dish reference*			
	Contract date			
	Expiry date			
	Quantity contracted			
	Contract price			

VOLUMETRICS

ENTITY SIZE	
No. OF OCCURRENCES	
TOTAL	

Fig. 13.10

ENTITY DESCRIPTION

SYSTEM: *SKC SALES & ACCOUNTING*	DATE:
AUTHOR: *G. CUTTS*	PAGE: 2 of 9

NAME: *CREDIT*

NARRATIVE:

Key	Data item	Format	Len.	Comment
✓	Customer number			
✓	Invoice number			
✓	Credit note number			
	Credit note date			
	Credit value			

	VOLUMETRICS	
	ENTITY SIZE	
	No. OF OCCURRENCES	
	TOTAL	

Fig. 13.11

ENTITY DESCRIPTION

SYSTEM: *SKC SALES & ACCOUNTING*	DATE:
AUTHOR: *G. CUTTS*	PAGE: 3 of 9

NAME: *CUSTOMER*

NARRATIVE:

Key	Data item	Format	Len.	Comment
✓	Customer number Name Address (Delivery address)			

	VOLUMETRICS	
	ENTITY SIZE	
	No. OF OCCURRENCES	
	TOTAL	

Fig. 13.12

ENTITY DESCRIPTION

SYSTEM: SKC - SALES & ACCOUNTING	DATE:
AUTHOR: G. CUTTS	PAGE: 4 of 9

NAME: DISH

NARRATIVE:

Key	Data item	Format	Len.	Comment
✓	Dish reference			
	Dish description			
	Standard price			
	Quantity produced			

	VOLUMETRICS	
	ENTITY SIZE	
	No. OF OCCURRENCES	
	TOTAL	

Fig. 13.13

```
┌─────────────────────────────────────────────────────────┐
│ ENTITY DESCRIPTION                                        │
├───────────────────────────────────┬───────────────────────┤
│ SYSTEM: SKC SALES & ACCOUNTING     │ DATE:                 │
├───────────────────────────────────┼───────────────────────┤
│ AUTHOR: G. CUTTS                   │ PAGE: 5 of 9          │
└───────────────────────────────────┴───────────────────────┘
```

NAME: *INVOICE HEAD*

NARRATIVE:

Key	Data item	Format	Len.	Comment
✓	Customer number			
✓	Invoice number			
	Invoice date			
	Total invoice value			

	VOLUMETRICS	
	ENTITY SIZE	
	No. OF OCCURRENCES	
	TOTAL	

Fig. 13.14

ENTITY DESCRIPTION

SYSTEM: *SKC SALES & ACCOUNTING*	DATE:
AUTHOR: *G. CUTTS*	PAGE: *6* of *9*

NAME: *INVOICE LINE*

NARRATIVE:

Key	Data item	Format	Len.	Comment
✓	{ Customer number			
✓	{ Invoice number			
✓	Dish reference			
✓	{ Customer number			
✓	{ SKC order number			
	Quantity delivered			
	Invoice line value			
	(Contract number)*			

	VOLUMETRICS	
	ENTITY SIZE	
	No. OF OCCURRENCES	
	TOTAL	

Fig. 13.15

ENTITY DESCRIPTION

SYSTEM: *SKC-SALES & ACCOUNTING*	DATE:
AUTHOR: *G. CUTTS*	PAGE: 7 of 9

NAME: *ORDER HEAD*

NARRATIVE:

Key	Data item	Format	Len.	Comment
	{ Customer number } { SKC order number } (Order number) Date of order Order stopped			 Yes/No

	VOLUMETRICS	
	ENTITY SIZE	
	No. OF OCCURRENCES	
	TOTAL	

Fig. 13.16

ENTITY DESCRIPTION

SYSTEM: *SKC SALES & ACCOUNTING*	DATE:
AUTHOR: *G. CUTTS*	PAGE: *8* of *9*

NAME: *ORDER LINE*

NARRATIVE: *Not merged with invoice line since it is created first.*

Key	Data item	Format	Len.	Comment
✓ ✓ ✓	{ Customer number } { SKC order number } Dish reference Quantity ordered (Contract number)* Quantity to be delivered Quantity delivered			
✓ ✓	{ Customer number } { Invoice number }			

	VOLUMETRICS	
	ENTITY SIZE	
	No. OF OCCURRENCES	
	TOTAL	

Fig. 13.17

ENTITY DESCRIPTION

SYSTEM: SKC SALES & ACCOUNTING	DATE:
AUTHOR: G. CUTTS	PAGE: 9 of 9

NAME: PAYMENT

NARRATIVE:

Key	Data item	Format	Len.	Comment
✓ ✓ ✓	{ Customer number Invoice number Payment number }			
	Payment date Payment value			

VOLUMETRICS

ENTITY SIZE	
No. OF OCCURRENCES	
TOTAL	

Fig. 13.18

13.4 SKC: ENTITY MODEL, TASK 3.4

Figure 13.19 shows the entities produced by part 3. Each entity rectangle is annotated with its key. Foreign keys within the entity's attributes are listed below the lower horizontal line.

The invoice line entity key and the order line entity key are identical. The entity descriptions could have been merged during the previous task. However, since order line entity occurrences exist in the system before invoice line entity occurrences, merging was not appropriate. A one-to-one relationship exists between order line and invoice line. Order line is a member of the order head-order line relationship, the dish-order line relationship, and the invoice head-order line relationship. There is no relationship between invoice head and invoice line except via the order line. The relationships are shown in Figure 13.20.

The final step is to mark the dominant part of each composite key as a foreign key and to connect these and the genuine foreign keys. Note that the foreign keys in the contract entity were marked as optional attributes. Figure 13.21 shows the entity model derived from the third normal form data structures.

13.5 SKC: LOGICAL ENTITY MODEL, TASK 3.5

The logical entity model derived from the entity model, part 2, Figure 12.4, and the entity model, part 3, Figure 13.21, is shown in Figure 13.22. The nine common entities and the order line status are retained. The order line status represents a specific user requirement. All of the relationships from the part 2 model are retained. There are three additional relationships on the part 3 model. They are:

Invoice head to order line This relationship is not carried across to the logical entity model since the relationship is correctly from invoice line to order line.

Invoice line to order line This relationship is retained. Each order line results in one delivery which results in one invoice line. The relationship is one to one.

Invoice line to contract This relationship is not carried across to the logical entity model as a relationship exists via the order line entity.

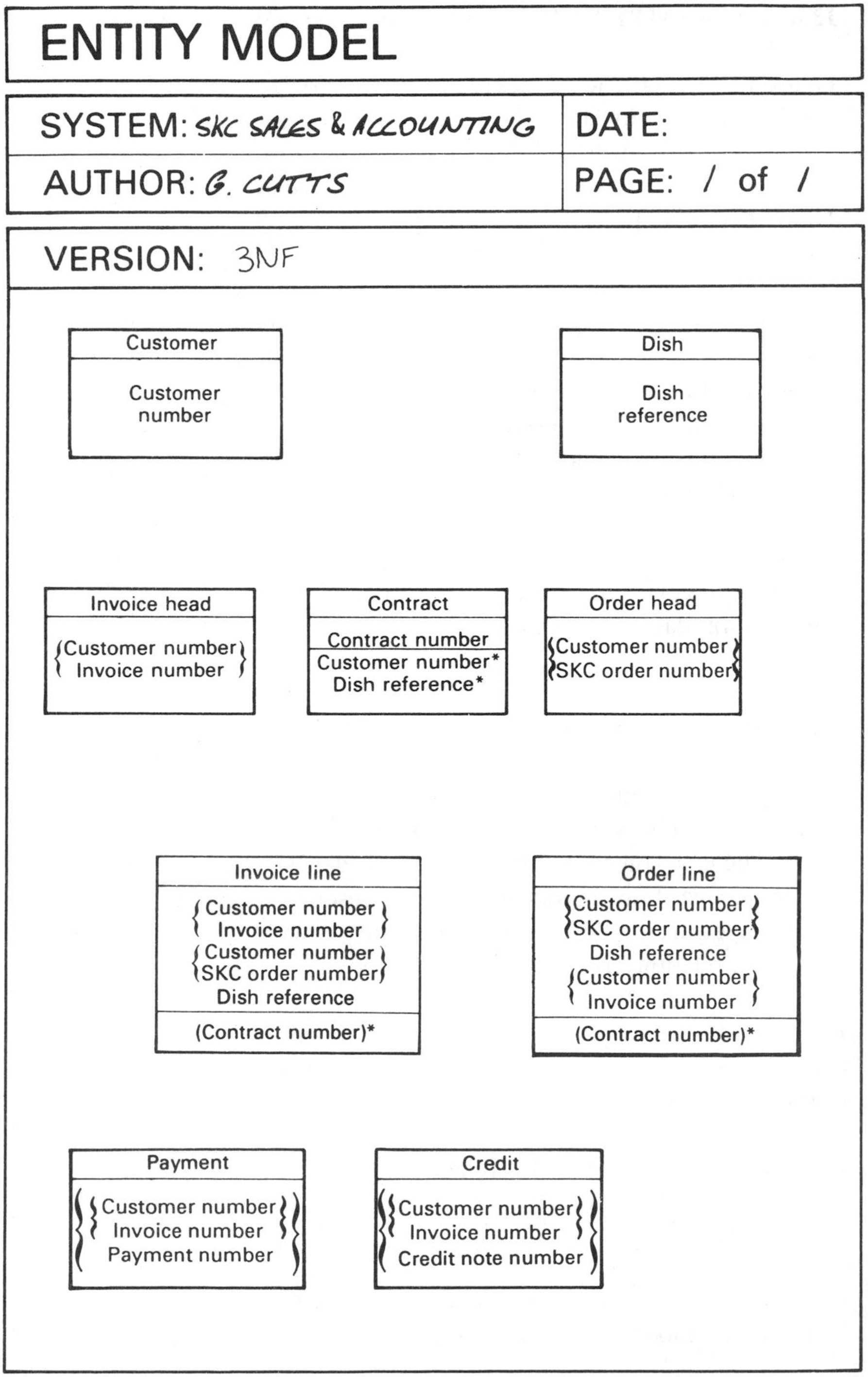

ENTITY MODEL

SYSTEM: *SKC SALES & ACCOUNTING*　DATE:

AUTHOR: *G. CUTTS*　PAGE: / of /

VERSION: 3NF

Customer
Customer number

Dish
Dish reference

Invoice head
{ Customer number } { Invoice number }

Contract
Contract number Customer number* Dish reference*

Order head
{ Customer number } { SKC order number }

Invoice line
{ Customer number } { Invoice number } { Customer number } { SKC order number } Dish reference
(Contract number)*

Order line
{ Customer number } { SKC order number } Dish reference { Customer number } { Invoice number }
(Contract number)*

Payment
{ { Customer number } } { Invoice number } { Payment number }

Credit
{ { Customer number } } { Invoice number } { Credit note number }

Fig. 13.19

ENTITY MODEL

SYSTEM: *SKC SALES & ACCOUNTING*	DATE:
AUTHOR: *G. CUTTS*	PAGE: / of /

VERSION: 3NF

Compound keys connected

Customer

Customer
number

Dish

Dish
reference

Invoice head

{ Customer number }
{ Invoice number }

Contract

Contract number
Customer number*
Dish reference*

Order head

{ Customer number }
{ SKC order number }

Invoice line

{ Customer number }
{ Invoice number }
{ Customer number }
{ SKC order number }
Dish reference

(Contract number)*

Order line

{ Customer number }
{ SKC order number }
Dish reference
{ Customer number }
{ Invoice number }

(Contract number)*

Payment

{ Customer number }
Invoice number
Payment number

Credit

{ Customer number }
Invoice number
Credit note number

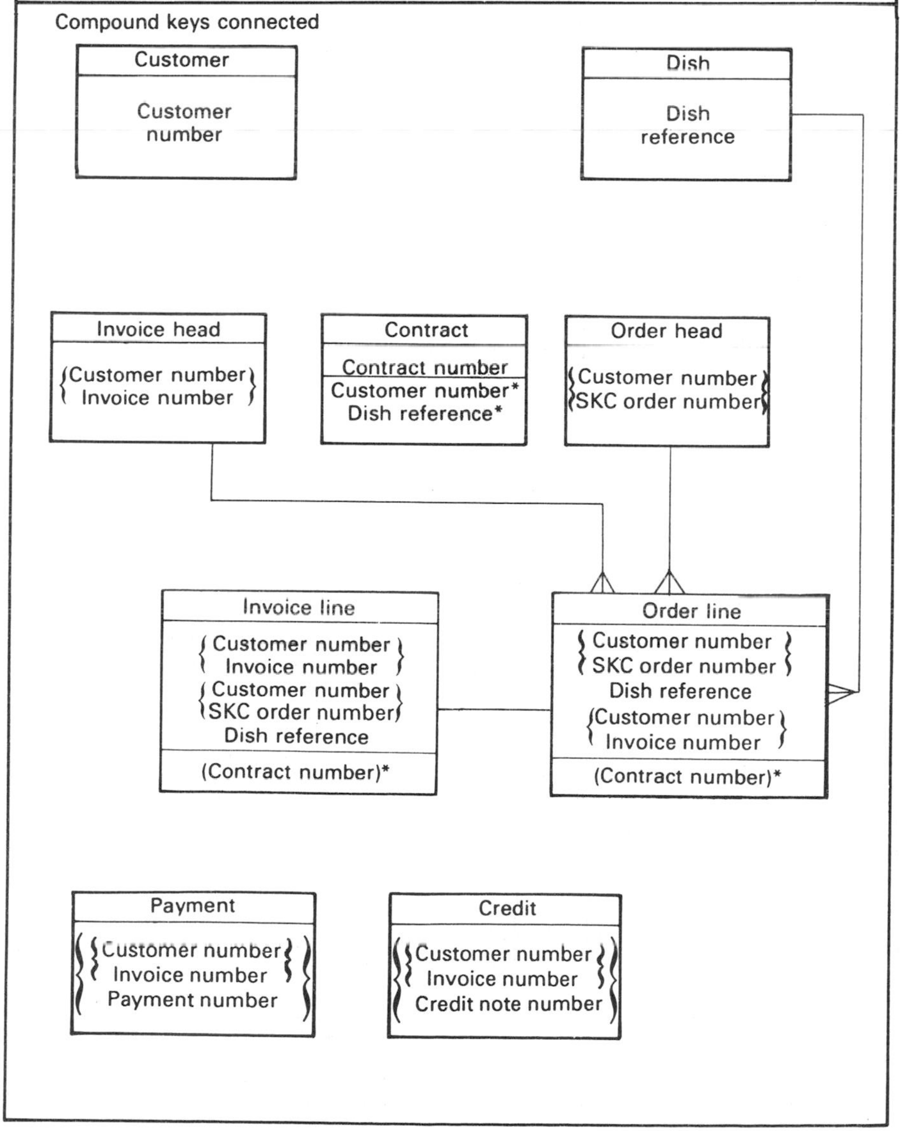

Fig. 13.20

ENTITY MODEL

SYSTEM: SKC SALES & ACCOUNTING   DATE:

AUTHOR: G. CUTTS   PAGE: 1 of 1

VERSION: 3NF

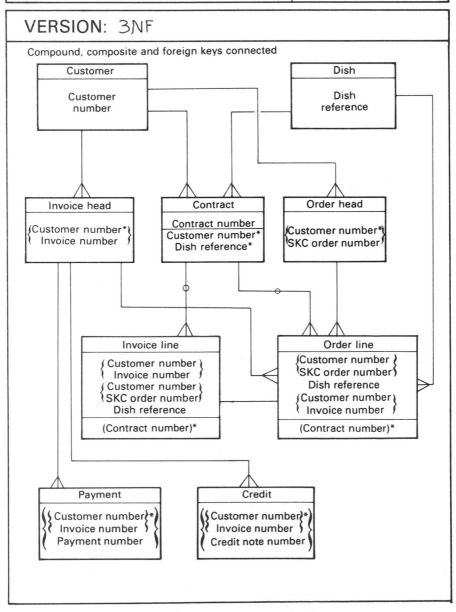

Fig. 13.21

ENTITY MODEL

SYSTEM: *SKC SALES & ACCOUNTING*	DATE:
AUTHOR: *G. CUTTS*	PAGE: *1* of *1*

VERSION: *LOGICAL*

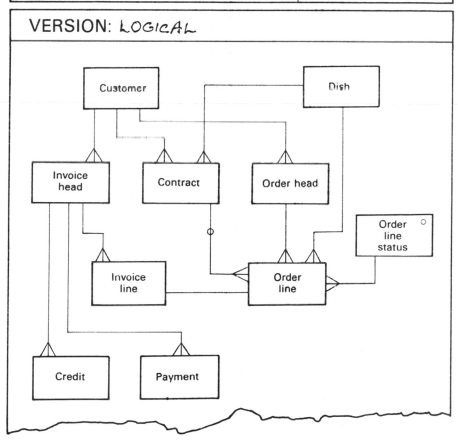

Fig. 13.22

13.6 SKC: LOGICAL ENTITY DESCRIPTIONS, TASK 3.6

The logical entity descriptions are derived from the entity descriptions from part 2, Figures 12.8 to 12.16, and the entity descriptions from part 3, Figures 13.10 to 13.18. They are shown in Figures 13.23 to 13.31.

   The logical entity descriptions were formed by simple merging of the attribute lists, with the exception of the following entity descriptions.

Dish   The quantity produced is used to update the quantity in stock. Quantity produced is deleted, quantity in stock is retained. There is no requirement to store the number of portions in a pack.

Invoice line The line number attribute is properly represented by the compound key made up of the composite key customer number/SKC order number and the simple key dish reference. The contract number is retained and, therefore, the contract price is not required.

Order head   The delivery address is not retained with the order as customers are only allowed one current delivery address. Delivery address is an attribute of the customer entity.

Order line   The contract number is retained and, therefore, the contract price is not required. The relationship to invoice head has been deleted, so the composite key customer number-invoice number is not required as an attribute.

ENTITY DESCRIPTION

| SYSTEM: *SKC SALES & ACCOUNTING* | DATE: |
| AUTHOR: *G. CUTTS* | PAGE: *1* of *9* |

| NAME: *CONTRACT* | *LOGICAL* |

NARRATIVE:

Key	Data item	Format	Len.	Comment
√	Contract number			
	Customer number*			
	Dish reference*			
	Contract date			
	Expiry date			
	Quantity contracted			
	Contract price			
	Quantity ordered to date			

	VOLUMETRICS	
	ENTITY SIZE	
	No. OF OCCURRENCES	
	TOTAL	

Fig. 13.23

ENTITY DESCRIPTION

SYSTEM: SKC SALES & ACCOUNTING	DATE:
AUTHOR: G. CUTTS	PAGE: 2 of 9

NAME: CREDIT NOTE	LOGICAL

NARRATIVE:

Key	Data item	Format	Len.	Comment
✓	Customer number			
✓	Invoice number			
✓	Credit note number			
	Credit note date			
	Credit value			

VOLUMETRICS

ENTITY SIZE	
No. OF OCCURRENCES	
TOTAL	

Fig. 13.24

ENTITY DESCRIPTION

SYSTEM: SKC SALES & ACCOUNTING	DATE:
AUTHOR: G. CUTTS	PAGE: 3 of 9

NAME: CUSTOMER	LOGICAL

NARRATIVE:

Key	Data item	Format	Len.	Comment
✓	Customer number			
	Name			
	Address			5 lines
	(Delivery address)			5 lines
	Stop listed			Yes/No
	Credit limit			£ value
	Credit time			Number of months
	Management clearance			Yes/No
	Activity			

	VOLUMETRICS	
	ENTITY SIZE	
	No. OF OCCURRENCES	
	TOTAL	

Fig. 13.25

ENTITY DESCRIPTION

SYSTEM: *SKC SALES & ACCOUNTING*	DATE:
AUTHOR: *G. CUTTS*	PAGE: *4* of *9*

NAME: *DISH*	*LOGICAL*

NARRATIVE:

Key	Data item	Format	Len.	Comment
✓	Dish reference			
	Dish description			
	Standard price			
	Quantity in stock			

	VOLUMETRICS	
	ENTITY SIZE	
	No. OF OCCURRENCES	
	TOTAL	

Fig. 13.26

ENTITY DESCRIPTION

SYSTEM: *SKC SALES & ACCOUNTING*	DATE:	
AUTHOR: *G. CUTTS*	PAGE: *5* of *9*	

NAME: *INVOICE HEAD*	*LOGICAL*

NARRATIVE:

Key	Data item	Format	Len.	Comment
	{ Customer number } { Invoice number }			
	Invoice date Total invoice value			

	VOLUMETRICS	
	ENTITY SIZE	
	No. OF OCCURRENCES	
	TOTAL	

Fig. 13.27

ENTITY DESCRIPTION

SYSTEM: *SKC SALES & ACCOUNTING*	DATE:
AUTHOR: *G. CUTTS*	PAGE: *6* of *9*

NAME: *INVOICE LINE*	*LOGICAL*

NARRATIVE:

Key	Data item	Format	Len.	Comment
✓ ✓ ✓ ✓	{ Customer number Invoice number SKC order number } Dish reference Quantity delivered (Standard price) Invoice line value (Contract number)*			Reduced from { Customer number } { Invoice number } { Customer number } { SKC order number }

	VOLUMETRICS	
	ENTITY SIZE	
	No. OF OCCURRENCES	
	TOTAL	

Fig. 13.28

ENTITY DESCRIPTION

SYSTEM: *SKC SALES & ACCOUNTING*	DATE:
AUTHOR: *G. CUTTS*	PAGE: 7 of 9

NAME: *ORDER HEAD*	*LOGICAL*

NARRATIVE:

Key	Data item	Format	Len.	Comment
✓ ✓	{ Customer number } { SKC order number } (Order number) Date of order Order stopped			Yes/No

	VOLUMETRICS	
	ENTITY SIZE	
	No. OF OCCURRENCES	
	TOTAL	

Fig. 13.29

ENTITY DESCRIPTION

SYSTEM: *SKC SALES & ACCOUNTING*	DATE:
AUTHOR: *G. CUTTS*	PAGE: *8* of *9*

NAME: *ORDER LINE*	*LOGICAL*

NARRATIVE:

Key	Data item	Format	Len.	Comment
✓ ✓ ✓	{Customer number} {SKC order number} Dish reference Quantity ordered (Contract number)* Quantity to be delivered Quantity delivered			
✓	Order line status			

	VOLUMETRICS	
	ENTITY SIZE	
	No. OF OCCURRENCES	
	TOTAL	

Fig. 13.30

ENTITY DESCRIPTION

SYSTEM: *SKC SALES & ACCOUNTING*	DATE:
AUTHOR: *G. CUTTS*	PAGE: *9* of *9*

NAME: *PAYMENT*	*LOGICAL*

NARRATIVE:

Key	Data item	Format	Len.	Comment
✓	Customer number			
✓	Invoice number			
✓	Payment number			
	Payment date			
	Payment value			
	Payment type			

	VOLUMETRICS		
	ENTITY SIZE		
	No. OF OCCURRENCES		
	TOTAL		

Fig. 13.31

Chapter 14
SKC Part 4: Logical Process Design

Part 3 did not introduce any new entities. The entity function matrix, entity life histories and function descriptions produced during part 2 are up to date. Task 4.1, the review of the stage A documentation, is not required.

14.1 SKC: THE LOGICAL PROCESS CATALOGUE, TASK 4.2

The primitive functions from the required physical DFDs have been posted to the logical process catalogue. The catalogue is shown in Figure 14.1.

14.2 SKC: LOGICAL PROCESS OUTLINES, TASK 4.3

Seventeen logical process outlines have been produced to correspond with the ten on-line entries and the seven batch mode entries in the logical process catalogue. The logical process outlines are shown in Figures 14.2 to 14.20.

LOGICAL PROCESS CATALOGUE

SYSTEM: SKC	DATE:
AUTHOR: G.CUTTS	PAGE: 1 of 1

Mode	Process no	DFD nos	Process name
On-line			
	1	1.1	Insert contract details
	2	1.2	Insert sales order details
	3	1.7	Update dish stock
	4	1.8	Sales order cancellation
	5	1.9	Sales order rejection
	6	1.10	Sales order special acceptance
	7	2.1	Record a customer's payment
	8	2.2	Insert account details
	9	2.6	Insert invoice line detils
	10	4.1	Maintain dish details
Batch daily			
	11	1.3	Produce order acknowledgment
		1.5	
	12	1.4	Produce despatch note
		1.6	
Batch weekly			
	13	2.3	Produce invoices and credit notes
		2.4	
		2.8	
	14	3	Produce management report
Batch monthly			
	15	2.7	Produce sales statement
	16	4.3	Archive customers and contracts
	17	4.2	Archive sales orders

Fig. 14.1

LOGICAL PROCESS OUTLINE

SYSTEM: *SKC SALES & ACCOUNTING*	DATE:
AUTHOR: *G. CUTTS*	PAGE: *1* of *19*

PROCESS No: *1*   NAME: *Insert contract details*

MODE: *On-line*   FREQUENCY: VOLUME:

BRIEF DESCRIPTION: *Contract details are inserted into the database on completion of successful negotiation.*
DFD Functions. *1.1*

ENTITY NAME	Contract										
EFFECT	I										
VALID PREV.	—										
SET TO	1										

Op. no.	Entity Name	Effect	Status Ind. Valid prev.	Set to	Description Narrative	Ref.	I/O ref.	Error ref.
1					Display screen format		F1	
2					Accept & validate the entries.	V1		E1
3	Contract	I	—	1	Insert the contract details.			

Fig. 14.2

LOGICAL PROCESS OUTLINE

SYSTEM: *SKC SALES & ACCOUNTING*	DATE:
AUTHOR: *G. CUTTS*	PAGE: 2 of *19*

PROCESS No: *2*　　NAME: *INSERT SALES ORDER DETAILS*

MODE: *On-line*　　FREQUENCY:　　VOLUME:

BRIEF DESCRIPTION: *Sales orders and inserted*
into the system.

DFD Functions. *1·2*

ENTITY NAME	Contract	Order head	Order line					
EFFECT	M	I	I					
VALID PREV.	1	—	—					
SET TO	1	1	1					

Op. no.	Entity Name	Effect	Status Ind. Valid prev.	Set to	Description Narrative	Ref.	I/O ref.	Error ref.
1					Display screen format.		F2	
2					Accept & validate the entries.	V2		E2
3	Order head	I	—	1	Insert the order head details.			
					Repeat 4 for all dish references on the order.			
4.1	Order line	I	—	1	Insert the order line details.			
4.2	Contract	M	1	1	If the order line contains a contract price obtain the relevant contract and add order line. quantity ordered to contract. quantity contracted.			

Fig. 14.3

LOGICAL PROCESS OUTLINE

SYSTEM: *SKC SALES & ACCOUNTING*	DATE:
AUTHOR: *G. CUTTS*	PAGE: *3* of *19*

PROCESS No: *3*   NAME: *UPDATE DISH STOCK*
MODE: *On-line*   FREQUENCY: VOLUME:

BRIEF DESCRIPTION: *Quantities of dishes produced are used to update the stock quantity available for sale.*

DFD Functions. *1.7*

ENTITY NAME	Dish									
EFFECT	M									
VALID PREV.	1									
SET TO	1									

Op. no.	Entity Name	Entity Effect	Status Ind. Valid prev.	Status Ind. Set to	Description Narrative	Ref.	I/O ref.	Error ref.
1					Display screen format.		F3	
2					Accept & validate the entries.	V3		E3
3	Dish	M	1	1	Add the dish stock. quantity produced to dish. quantity in stock.			

Fig. 14.4

LOGICAL PROCESS OUTLINE

SYSTEM: *SKC-SALES & ACCOUNTING*	DATE:
AUTHOR: *G. CUTTS*	PAGE: *4* of *19*

PROCESS No: *4*   NAME: *SALES ORDER CANCELLATION*
MODE: *On-Line*   FREQUENCY: VOLUME:
BRIEF DESCRIPTION: *The processing required following*
DFD Functions. *1·8*   *an order cancellation.*

ENTITY NAME	Dish	Order head	Order line				
EFFECT	M	D	D				
VALID PREV.	1	1,2,3,	1,2,3,				
SET TO	1	—	—				

Op. no.	Entity Name	Effect	Status Ind. Valid prev.	Set to	Description Narrative	Ref.	I/O ref.	Error ref.
1					Display screen format.		F4	
2					Accept and validate the entries.	V4		E2
3					Repeat 3 for all order lines on the order.			
3.1	Order line	R	—	—	If the order line status = 3 then operation 3.2.			
3.2	Dish	M	1	1	Add order line. quantity to be delivered to dish. quantity in stock.			
3.3	Order line	D	1,2,3	—	Delete the order line.			
4	Order head	D	1,2,3	—	Delete the order head.			

Fig. 14.5

LOGICAL PROCESS OUTLINE

SYSTEM: *SKC SALES & ACCOUNTING*	DATE:
AUTHOR: *G. CUTTS*	PAGE: *5* of *19*

PROCESS No: *5*   NAME: *SALES ORDER REJECTION*

MODE: *On-line*   FREQUENCY: VOLUME:

BRIEF DESCRIPTION: *The processing required following the rejection by SKC of a sales order.*

DFD Functions. *1·9*

ENTITY NAME	Order head	Order line							
EFFECT	D	D							
VALID PREV.	1	1							
SET TO	—	—							

Op. no.	Entity Name	Effect	Status Ind. Valid prev.	Set to	Description Narrative	Ref.	I/O ref.	Error ref.
1					Display screen format.		F5	
2					Accept & validate the entries.	V5		E5
3					Repeat 3 for all order lines on the order.			
3.1	Order line	D	1	—	Delete the order line.			
4	Order head	D	1	—	Deleter the order head.			

Fig. 14.6

LOGICAL PROCESS OUTLINE

SYSTEM: SKC SALES & ACCOUNTING	DATE:
AUTHOR: G. CUTTS	PAGE: 6 of 19

PROCESS No: 6 NAME: SALES ORDER SPECIAL ACCEPTANCE

MODE: On-line FREQUENCY: VOLUME:

BRIEF DESCRIPTION: The processing of a sales order after special acceptance by SKC management.

DFD Functions. 1.10

ENTITY NAME	Order head	Order line								
EFFECT	M	M								
VALID PREV.	1	1								
SET TO	2	2								

Op. no.	Entity Name	Entity Effect	Status Ind. Valid prev.	Status Ind. Set to	Description Narrative	Ref.	I/O ref.	Error ref.
1					Display screen format.		F6	
2					Accept & validate the input	V6		E6
3	Order head	M	1	2	Set order head. Order stopped = no.			
4	Order line	M	1	2	For all order lines on the order.			

Fig. 14.7

LOGICAL PROCESS OUTLINE

SYSTEM: *SKC SALES & ACCOUNTING*	DATE:
AUTHOR: *G. CUTTS*	PAGE: 7 of *19*

PROCESS No: *7*   NAME: *RECORD A CUSTOMERS PAYMENT*

MODE: *On-line*   FREQUENCY: VOLUME:

BRIEF DESCRIPTION: *Payments from customers are recorded against invoices.*

DFD Functions: *2·1*

ENTITY NAME	Invoice head	Payment							
EFFECT	R	I							
VALID PREV.	—	—							
SET TO	—	1							

Op. no.	Entity Name	Effect	Status Ind. Valid prev.	Set to	Description Narrative	Ref.	I/O ref.	Error ref.
1					Display screen format.		F7	
2					Accept & validate the input.	V7		E7
3					Repeat 3 until the input (3.2) = A (allocate).			
3.1	Invoice head	R	—	—	Read & display the details of the first/next invoice head.		F8	
3.2					Accept & validate the input.	V8		E8
4	Payment	I	—	1	Insert the payment details.			

Fig. 14.8

LOGICAL PROCESS OUTLINE

SYSTEM: *SKC SALES & ACCOUNTING*	DATE:
AUTHOR: *G. CUTTS*	PAGE: *8* of *19*

PROCESS No: *8*   NAME: *INSERT ACCOUNT DETAILS*

MODE: *On-line*   FREQUENCY: VOLUME:

BRIEF DESCRIPTION: *Account details are inserted into the database following authorisating of the customer's details.*

DFD Functions. *2·2*

ENTITY NAME	Customer										
EFFECT	I										
VALID PREV.	—										
SET TO	1										

Op. no.	Entity Name	Effect	Status Ind. Valid prev.	Set to	Description Narrative	Ref.	I/O ref.	Error ref.
1					Display the screen format.		F9	
2					Accept & validate the input.	V9		E9
3	Customer	I	—	1	Insert the customer details.			

Fig. 14.9

LOGICAL PROCESS OUTLINE

SYSTEM: *SKC SALES & ACCOUNTING*	DATE:
AUTHOR: *G. CUTTS*	PAGE: *9* of *19*

PROCESS No: *9*   NAME: *INSERT INVOICE LINE DETAILS*

MODE: *On-line*   FREQUENCY: VOLUME:

BRIEF DESCRIPTION: *For cash delivery an invoice line is inserted into the database.*

DFD Functions. *2·6*

ENTITY NAME	Contract	Dish	Invoice line				
EFFECT	R	R	I				
VALID PREV.	–	–	–				
SET TO	–	–	1				

Op. no.	Entity Name	Effect	Status Ind. Valid prev.	Set to	Description Narrative	Ref.	I/O ref.	Error ref.
1					Display screen format		F10	
2					Accept and validate the input	V10		E10
					If delivery. contract delivery = yes then operation 3 else operation 4.			
3	Contract	R			Set invoice line. price each = contract. contract price.			
	Dish	R			Set invoice line. price each = dish. standard price.			
	Invoice line	I	–	1	Insert the invoice line details.			

Fig. 14.10

LOGICAL PROCESS OUTLINE

SYSTEM: *SKC SALES & ACCOUNTING*	DATE:
AUTHOR: *G. CUTTS*	PAGE: *10* of *19*

PROCESS No: *10*   NAME: *MAINTAIN DISH DETAILS*

MODE: *On-line*   FREQUENCY: VOLUME:

BRIEF DESCRIPTION:

DFD Functions. *4.1*   *All processing of dish static detail.*

ENTITY NAME	Dish	Dish	Dish					
EFFECT	I	M	D					
VALID PREV.	—	1	1					
SET TO	1	1						

Op no.	Entity Name	Effect	Status Ind. Valid prev.	Set to	Description Narrative	Ref.	I/O ref.	Error ref.
1					Display screen format.		F11	
2					Accept & validate the input.	V11		E11

					If input = I then 3
					    input = M then 4
					    input = D then 5			

3.1					Display screen format.		F12	
3.2					Accept & validate the input.	V12		E12
3.3	Dish	I	—	1	Insert the dish details.			
4.1	Dish	R			Read the dish entity referenced by the input & display screen format.		F13	
4.2					Accept & validate the input.	V13		E13
4.3	Dish	M	1	1	Insert the dish details to replace the current occurrence.			
5	Dish	D	1	—	Delete the dish details referenced by the input.			

Fig. 14.11

LOGICAL PROCESS OUTLINE

SYSTEM: *SKC SALES & ACCOUNTING*	DATE:
AUTHOR: *B. CUTTS*	PAGE: *11* of *19*

PROCESS No: *11*   NAME: *PRODUCE ORDER ACKNOWLEDGEMENT*

MODE: *BATCH*   FREQUENCY: *DAILY* VOLUME:

BRIEF DESCRIPTION: *All newly inserted sales orders are checked for credit violation. If OK an order*
DFD Functions. *1.3 & 1.5 acknowledgement is produced.*

ENTITY NAME	Customer	Order head	Order line					
EFFECT	R	M	M					
VALID PREV.	—	1	1					
SET TO	—	3	2					

Op. no.	Entity Name	Effect	Status Ind. Valid prev.	Set to	Description Narrative	Ref.	I/O ref.	Error ref.
					Repeat for all order head where order head. status = 1.			
1	Order head	R			Read the first/next order head.			
2	Customer	R			Read the customer given by order head. customer number.			
3	Order head	M	1	1	If the customer. stop listed = yes then set order head. order stopped = yes : END.			
4					If the customer. stop listed = no then set order head. order stopped = no and operations 5 & 6			
5	Order line	M	1	2	Produce an order acknowledgement containing all of the order lines. Modify the order lines status.		O19	
6	Order head	M	1	3	Modify the order head. status.			

Fig. 14.12

LOGICAL PROCESS OUTLINE

SYSTEM: *SKC SALES & ACCOUNTING*	DATE:
AUTHOR: *G. CUTTS*	PAGE: *12* of *19*

PROCESS No: *12*   NAME: *PRODUCE DESPATCH NOTE*

MODE: *BATCH*   FREQUENCY: *DAILY* VOLUME:

BRIEF DESCRIPTION: *All accepted sales orders are checked with regard to stock availability. For all order lines with sufficient stock available then a despatch note is produced.*

DFD Functions *14, 16*

ENTITY NAME	Dish	Order line	Order head	Customer		
EFFECT	M	M	R	R		
VALID PREV.	1	2	—	—		
SET TO	1	4	—	—		

Op. no.	Entity Name	Effect	Status Ind. Valid prev.	Set to	Description Narrative	Ref.	I/O ref.	Error ref.
1	Order head	R			Repeat all operations for all order head entity occurrences. Read the first/next order head entity occurrence.			
2					If the order head. order stopped = yes then ignore.			
3.1	Order line	R			Repeat 3 for every order with status 1 on the order head. Read the first/next order line,			
3.2	Dish	R			Read the dish referenced by order line.dish reference.			
3.3	Dish	M	1	1	If dish. quantity in stock $\geqslant$ order line. quantity ordered then decrement dish. quantity in stock.			
3.4	Customer	R			Produce a despatch note.			
3.5	Order line	M	2	4	Modify the order line status.		O11	

Fig. 14.13

LOGICAL PROCESS OUTLINE

SYSTEM: *SKC SALES & ACCOUNTING*	DATE:
AUTHOR: *G. CUTTS*	PAGE: *13* of *19*

PROCESS No: *13*   NAME: *PRODUCE INVOICES & CREDIT NOTES*
MODE: *BATCH*   FREQUENCY: *WEEKLY* VOLUME:
BRIEF DESCRIPTION: *All invoices & credit notes not previously produced should be printed. Customer credit limits are checked.*
DFD Functions. *2·3, 2·4, 2·8*

ENTITY NAME	Credit note	Customer	Invoice head	Invoice line	Order line	
EFFECT	I	M	I	M	M	
VALID PREV.	—	1	—	1	4	
SET TO	1	1	1	2	5	

Op. no.	Entity Name	Effect	Status Ind. Valid prev.	Status Ind. Set to	Description Narrative	Ref.	I/O ref.	Error ref.
					Repeat all operations for all customers.			
·1	Customer	R	—	—	Read the first/next customer.			
2	Credit note	I	—	1	If a credit authorisation exists for this customer insert the credit note into the system & produce a credit note.		I10 O12	
					For all invoice lines. status = 1			
3.1	Invoice head	I	—	1	Insert an invoice head with invoice head. invoice number incremented by 1.			
3.2					Print a sales invoice.		O13	
3.3	Invoice line	M	1	2	Modify all invoice lines.			
					Where invoice lines. invoice number = invoice head. invoice number.			
3.4	Order line	M	4	5	Modify all order lines.			

Fig. 14.14

LOGICAL PROCESS OUTLINE

SYSTEM: *SKC SALES & ACCOUNTING*	DATE:
AUTHOR: *G. CUTTS*	PAGE: *14* of *19*

PROCESS No: *13*   NAME: *PRODUCE INVOICES & CREDIT NOTES*
MODE: *BATCH*   FREQUENCY: *WEEKLY* VOLUME:
BRIEF DESCRIPTION: *All invoices & credit notes not previosly produced should be printed. Customer* DFD Functions. *2-2, 2-4, 2-8   credit limits are checked.*

ENTITY NAME	Credit note	Customer	Invoice head	Invoice line	Order line	
EFFECT	I	M	I	M	M	
VALID PREV.	—	1	—	1	4	
SET TO	1	1	1	2	5	

Op. no.	Entity Name	Effect	Status Ind. Valid prev.	Set to	Description Narrative	Ref.	I/O ref.	Error ref.
4.1	Invoice head	R	—	—	Read all the invoice heads, payment & credit notes.			
4.2	Payment	R	—	—				
4.3	Credit note	R	—	—				
4.4	Customer	R	—	—	Check if management clearance exists for the customer. (i.e. customer. management clearance = yes.)			
4.5	Customer	M	1	1	Modify the customer. status. If there are no invoice head inserts, add 1 to customer. activity else set customer. activity = 0	DT1		

Fig. 14.15

| LOGICAL PROCESS OUTLINE | | | | |

SYSTEM: *SKC SALES & ACCOUNTING*		DATE:		
AUTHOR: *G. CUTTS*		PAGE: *15* of *19*		

PROCESS No: *14*   NAME: *PRODUCE MANAGEMENT REPORT*

MODE: *BATCH*   FREQUENCY: *WEEKLY* VOLUME:

BRIEF DESCRIPTION: *By reference to orders, invoices and*
DFD Functions: *3*   *contracts produce the weekly management report.*

ENTITY NAME					
EFFECT					
VALID PREV.					
SET TO					

Op. no.	Entity Name	Effect	Status Ind. Valid prev.	Set to	Description Narrative	Ref.	I/O ref.	Error ref.
					Repeat all operations for all customers.			
1	Customer	R			Read the first/next customer.			
					Repeat 2 for all order heads.			
2.1	Order head	R	—	—	Read the order head.			
2.2					If order head. order stopped = yes then produce a line on the report else repeat 3 for all order lines.		020	
3.1	Order line	R	—	—	Read the order line.			
3.2	Invoice line	R	—	—	Read the invoice line & contract for the order line.			
3.3	Contract	R	—	—				
3.4					Produce a line on the report.		021	

Fig. 14.16

LOGICAL PROCESS OUTLINE

SYSTEM: SKC SALES & ACCOUNTING	DATE:
AUTHOR: G. CUTTS	PAGE: 16 of 19

PROCESS No: 15   NAME: PRODUCE SALES STATEMENT

MODE: BATCH   FREQUENCY: MONTHLY VOLUME:

BRIEF DESCRIPTION: For each active customer produce a sales statement.

DFD Functions. 2.7

ENTITY NAME	Credit note	Credit note	Customer	Invoice head	Invoice head	Invoice line	Payment	Payment
EFFECT	R	A	R	M	A	A	R	A
VALID PREV.	—	1	—	1	1	2	—	1
SET TO	—	—	—	1	—	—	—	—

Op. no.	Entity Name	Effect	Status Ind. Valid prev.	Set to	Description / Narrative	Ref.	I/O ref.	Error ref.
					Repeat all operations for each customer.			
1	Customer	R	—	—	Read the customer & produce a statement head.		O30	
					Repeat 2 for each invoice head.			
2.1	Invoice head	R			Read the first/next invoice head.			
2.2	Payment	R			Read all the payments for the invoice head.			
2.3	Credit note	R			Read all the credit notes for the invoice head.			
2.4					If invoice head. total invoice value ≤ total (payment. payment value + credit note. credit value)			
					then operation 4, continue.			
2.5					Produce a statement line for invoice head payment credit note		O31 O32 O33	
3					Produce a statement trailer.		O34	
					continued			

Fig. 14.17

LOGICAL PROCESS OUTLINE

SYSTEM: *SKC SALES & ACCOUNTING*	DATE:
AUTHOR: *G. CUTTS*	PAGE: *17* of *19*

PROCESS No: *15*   NAME: *PRODUCE SALES STATEMENT*

MODE: *BATCH*   FREQUENCY: *MONTHLY* VOLUME:

BRIEF DESCRIPTION: *For each active customer*
DFD Functions. *2·7*   *produce a sales statements.*

ENTITY NAME	Credit note	Credit note	Customer	Invoice head	Invoice head	Invoice line	Payment	Payment
EFFECT	R	A	R	M	A	A	R	A
VALID PREV.	—	1	—	1	1	2	—	1
SET TO	—	—	—	1	—	—	—	—

Op. no.	Entity Name	Effect	Status Ind. Valid prev.	Set to	Description Narrative	Ref.	I/O ref.	Error ref.
4.1	Invoice line	A	2	—	Archive all invoice line for the invoice head.			
4.2	Payment	A	1	—	Archive all payment for the invoice head.			
4.3	Credit note	A	1	—	Archive all credit note for the invoice head.			
4.4	Invoice head	A	1	—	Archive the invoice head. Return.			

Fig. 14.18

LOGICAL PROCESS OUTLINE

SYSTEM: *Skc Sales & Accounting*	DATE:
AUTHOR: *G. Cutts*	PAGE: *18* of *19*

PROCESS No: *16*   NAME: *Archive Customers & Contracts*

MODE: *Batch*   FREQUENCY: *Monthly* VOLUME:

BRIEF DESCRIPTION: *For all non-active customers and completed contracts copy the details from the database to the archive files.*

DFD Functions. 4·3

ENTITY NAME	Contract	Customer						
EFFECT	A	A						
VALID PREV.	1	1						
SET TO	—	—						

Op. no.	Entity Name	Effect	Status Ind. Valid prev.	Set to	Description Narrative	Ref.	I/O ref.	Error ref.
					Repeat all processing for each customer.			
					Repeat 1 for all contracts for the customer.			
1	Customer	A	1	—	If there are no contracts and customer. activity ≥ 26 then archive the customer.			
2	Contract	A	1	—	If contract. quantity ordered to date ≥ contract. quality contracted then archive the contract.			

Fig. 14.19

LOGICAL PROCESS OUTLINE

SYSTEM: *SKC SALES & ACCOUNTING*	DATE:
AUTHOR: *G. CUTTS*	PAGE: *19* of *19*

PROCESS No: *17*   NAME: *ARCHIVE SALES ORDERS*

MODE: *BATCH*   FREQUENCY: *MONTHLY* VOLUME:

BRIEF DESCRIPTION: *For all orders where all odd lines have*
DFD Functions: *4·2*   *been invoiced, copy the details to the archive files.*

ENTITY NAME	Order head	Order line							
EFFECT	A	A							
VALID PREV.	3	4							
SET TO	—	—							

Op. no.	Entity Name	Entity Effect	Status Ind. Valid prev.	Status Ind. Set to	Description Narrative	Ref.	I/O ref.	Error ref.
1					Repeat for all order head. Read first/next order head. Repeat 2 for all order lines invoiced: = 0; order lines: = 0			
2.1	Order line	R			Read first/next order line.			
2.2					If order line. status = 4 then invoiced = invoiced + 1			
2.3					Order lines = order lines + 1 If order lines = invoiced then 3			
3.1	Order line	A	4	—	Archive all order lines.			
3.2	Order head	A	3	—	Archive the order head.			

Fig. 14.20

Appendix A
SSADM Standard Forms

Figure

structured systems analysis and design	STANDARD FORM

1 DOCUMENT FLOW DIAGRAM

SYSTEM:	DATE:
AUTHOR:	PAGE: of

GEOFF CUTTS – 1991

Fig. A1

structured systems analysis and design   STANDARD FORM

2 DATA FLOW DIAGRAM

SYSTEM:	DATE:
AUTHOR:	PAGE: of

LEVEL: 1	CURRENT/REQ.	PHYS./LOGICAL

GEOFF CUTTS   1991

Fig. A2

structured systems analysis and design   STANDARD FORM

3 DATA FLOW DIAGRAM

SYSTEM:	DATE:
AUTHOR:	PAGE: of

LEVEL:	CURRENT/REQ.	PHYS./LOGICAL
TITLE:		

GEOFF CUTTS – 1991

Fig. A3

structured systems analysis and design	STANDARD FORM

4 ENTITY MODEL

SYSTEM:	DATE:
AUTHOR:	PAGE: of

VERSION:

GEOFF CUTTS – 1991

Fig. A4

structured systems analysis and design	STANDARD FORM

5 DATA STORE/ENTITY X REF.

SYSTEM:	DATE:
AUTHOR:	PAGE: of

GEOFF CUTTS – 1991

Fig. A5

structured systems analysis and design   STANDARD FORM

6 PROBLEMS/REQUIREMENTS LIST

SYSTEM: DATE:

AUTHOR: PAGE: of

No.	Problem/requirement	Init.	Solution Reference

GEOFF CUTTS – 1991

Fig. A6

structured systems analysis and design	STANDARD FORM

7 ENTITY DESCRIPTION

SYSTEM:	DATE:
AUTHOR:	PAGE: of

NAME:	VERSION:

NARRATIVE:

Key	Data item	Comment
		GEOFF CUTTS – 1991

Fig. A7

structured systems analysis and design STANDARD FORM

8 INPUT OR OUTPUT DESCRIPTION

SYSTEM:	DATE:
AUTHOR:	PAGE: of

NARRATIVE:

Key	Data item	Comment

GEOFF CUTTS – 1991

Fig. A8

structured systems analysis and design	STANDARD FORM

9 DATA DICTIONARY

SYSTEM:	DATE:
AUTHOR:	PAGE: of:

DATA ITEM (user name):

SHORT NAME:

Type:

Length:

Format:

Characteristics

SIGNED: YES/NO
ROUNDED: YES/NO
PACKED: YES/NO
JUSTIFIED: YES/NO

Range:

Description:

GEOFF CUTTS – 1991

Fig. A9

10 ON-LINE DIALOGUE SPECIFICATION

SYSTEM:	DATE:
AUTHOR:	PAGE: of:

GEOFF CUTTS – 1991

Fig. A10

| structured systems analysis and design | STANDARD FORM |

11 LOGICAL FUNCTION DESCRIPTION

| SYSTEM: | DATE: |
| AUTHOR: | PAGE: of |

DFD Ref.	Name	Description

GEOFF CUTTS – 1991

Fig. A11

structured systems analysis and design   STANDARD FORM

12 ENTITY/FUNCTION MATRIX

SYSTEM:	DATE:
AUTHOR:	PAGE: of

GEOFF CUTTS – 1991

Fig. A12

13 ENTITY LIFE HISTORY

SYSTEM:	DATE:
AUTHOR:	PAGE: of

GEOFF CUTTS – 1991

Fig. A13

structured systems analysis and design	STANDARD FORM

14 NORMALISATION

SYSTEM:	DATE:
AUTHOR:	PAGE: of

DATA STRUCTURE:

UNF	1NF

GEOFF CUTTS   1991

Fig. A14(a)

structured systems analysis and design	STANDARD FORM

14 NORMALISATION

SYSTEM:	DATE:
AUTHOR:	PAGE: of

DATA STRUCTURE:

2NF	3NF

GEOFF CUTTS – 1991

Fig. A14(b)

structured systems analysis and design	STANDARD FORM

15 LOGICAL PROCESS CATALOGUE

SYSTEM:	DATE:
AUTHOR:	PAGE: of

Mode	Process no.	DFD nos.	Process name
			GEOFF CUTTS – 1991

Fig. A15

structured systems analysis and design   STANDARD FORM

16 LOGICAL PROCESS OUTLINE

SYSTEM: DATE:

AUTHOR: PAGE: of

PROCESS No: PROCESS NAME:
MODE: DFD Nos:
BRIEF DESCRIPTION:

ENTITY NAME										
EFFECT										
VALID PREV.										
SET TO										

Op. no.	Entity Name	Effect	Status Ind. VP	ST	Operation description	Ref.	I/O ref.	Error ref.

GEOFF CUTTS – 1991

Fig. A16(a)

structured systems analysis and design STANDARD FORM

16 LOGICAL PROCESS OUTLINE

SYSTEM: DATE:

AUTHOR: PAGE: of:

PROCESS NO: PROCESS NAME:

Op. no.	Entity Name	Effect	Status Ind. VP	ST	Operation description	Ref	I/O ref.	Error ref.

GEOFF CUTTS – 1991

Fig. A16(b)

structured systems analysis and design   STANDARD FORM

17 PHYSICAL ACCESS PATHS

SYSTEM:	DATE:
AUTHOR:	PAGE: of

PROCESS No:	PROCESS NAME:

Operation no.	Entity name	Effect	Access path

GEOFF CUTTS – 1991

Fig. A17

Appendix B
SSADM Version 4

The CCTA provided details of version 4 of SSADM for inclusion in this book. I am grateful to the CCTA and to NCC Blackwell, who publish the SSADM version 4 manuals,* for permission to reproduce the schematics.

Figure

B1 SSADM life cycle
B2 Stage 1 – Investigation of current environment
B3 Stage 2 – Business system options
B4 Stage 3 – Definition of requirements
B5 Stage 4 – Technical system options
B6 Stage 5 – Logical design
B7 Stage 6 – Physical design

*

Longworth, G. and Nicholls, D. (1987) *SSADM Manual. Volume 1: Tasks and Terms*. NCC Blackwell, Oxford.
Longworth, G. and Nicholls, D. (1987) *SSADM Manual. Volume 2: Techniques and Documentation*. NCC Blackwell, Oxford.

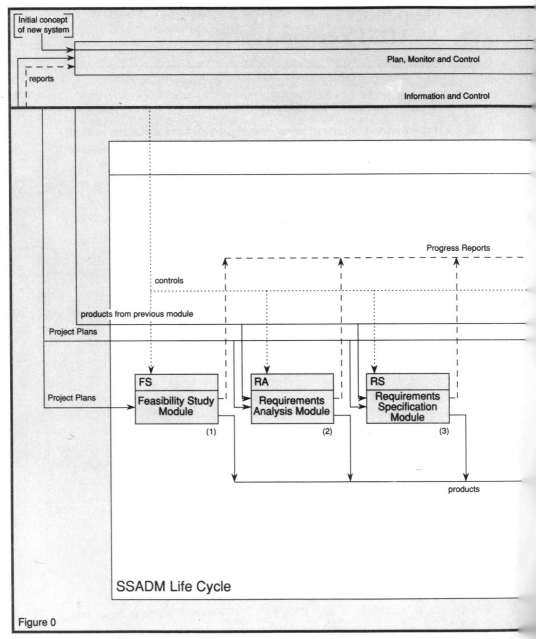

Figure 0

Fig. B1 SSADM life cycle

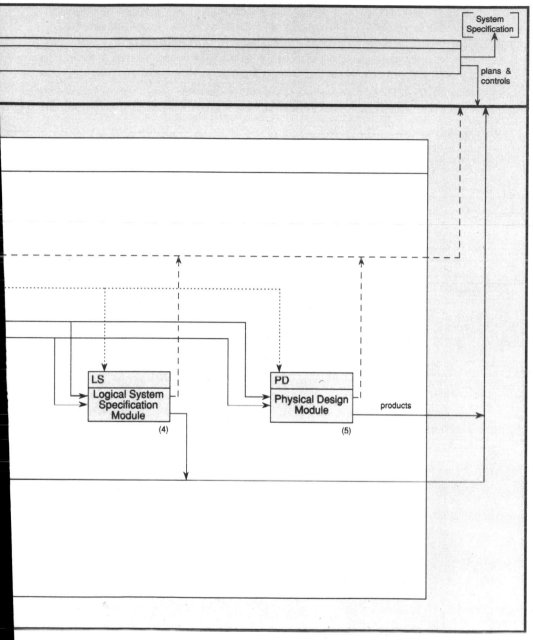

System
Specification

plans &
controls

LS
Logical System
Specification
Module
(4)

PD
Physical Design
Module
(5)

products

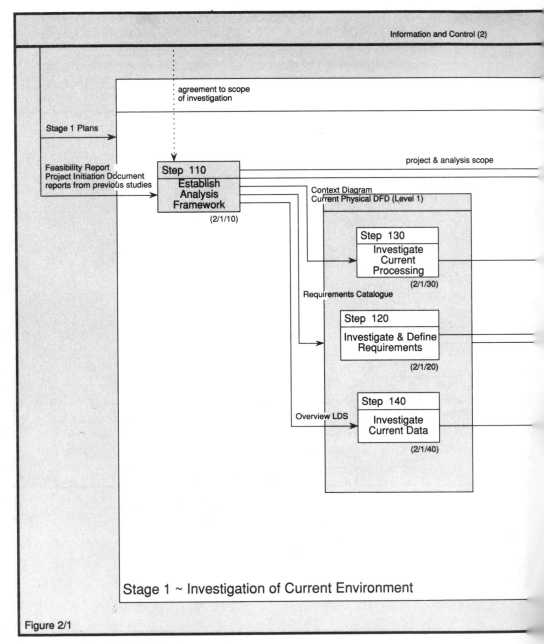

Stage 1 ~ Investigation of Current Environment

Figure 2/1

Fig. B2 Stage 1 – Investigation of current environment

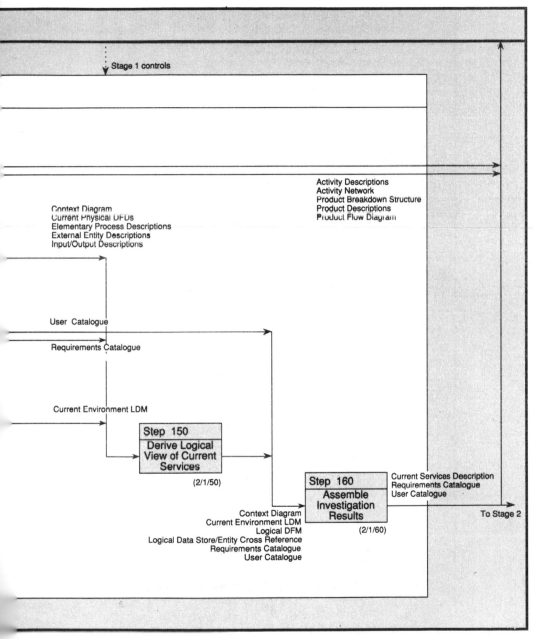

Stage 1 controls

Activity Descriptions
Activity Network
Product Breakdown Structure
Product Descriptions
Product Flow Diagram

Context Diagram
Current Physical DFDs
Elementary Process Descriptions
External Entity Descriptions
Input/Output Descriptions

User Catalogue

Requirements Catalogue

Current Environment LDM

Step 150
Derive Logical
View of Current
Services

(2/1/50)

Step 160
Assemble
Investigation
Results

(2/1/60)

Current Services Description
Requirements Catalogue
User Catalogue

To Stage 2

Context Diagram
Current Environment LDM
Logical DFM
Logical Data Store/Entity Cross Reference
Requirements Catalogue
User Catalogue

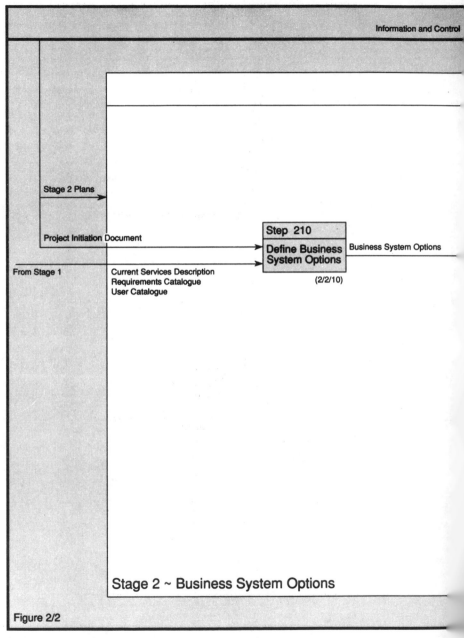

Information and Control

Stage 2 Plans

Project Initiation Document

From Stage 1

Current Services Description
Requirements Catalogue
User Catalogue

Step 210

Define Business
System Options

(2/2/10)

Business System Options

Stage 2 ~ Business System Options

Figure 2/2

Fig. B3 Stage 2 – Business system options

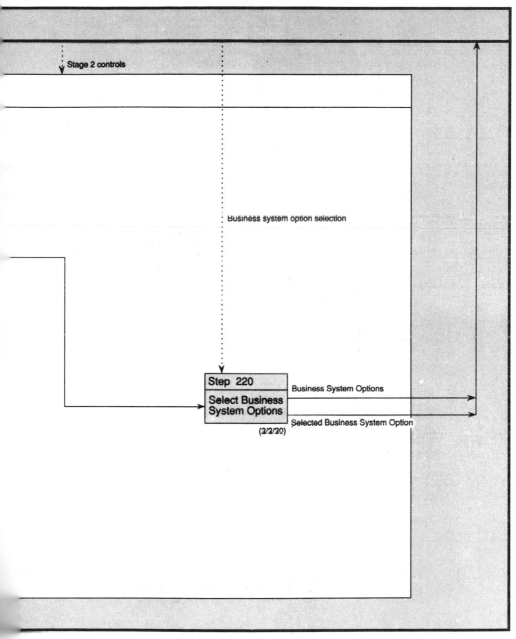

Stage 2 controls

Business system option selection

Step 220

Select Business
System Options

(2/2/20)

Business System Options

Selected Business System Option

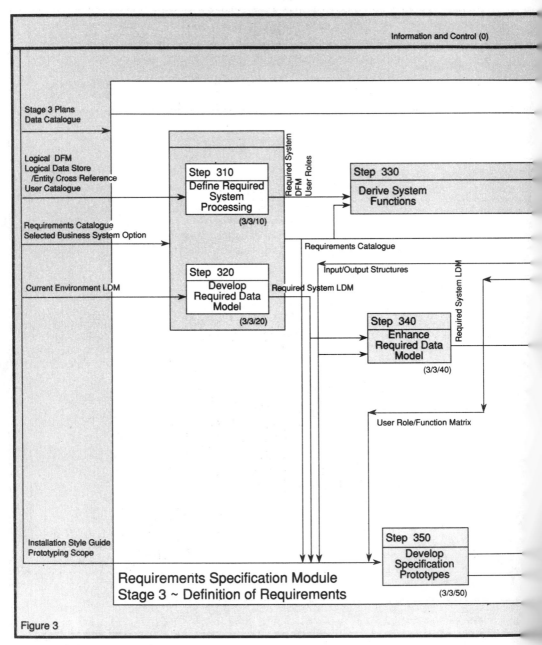

Information and Control (0)

Stage 3 Plans
Data Catalogue

Logical DFM
Logical Data Store
/Entity Cross Reference
User Catalogue

Requirements Catalogue
Selected Business System Option

Current Environment LDM

Installation Style Guide
Prototyping Scope

Required System DFM
User Roles

Step 310
Define Required
System
Processing
(3/3/10)

Step 320
Develop
Required Data
Model
(3/3/20)

Step 330
Derive System
Functions

Requirements Catalogue

Input/Output Structures

Required System LDM

Required System LDM

Step 340
Enhance
Required Data
Model
(3/3/40)

User Role/Function Matrix

Step 350
Develop
Specification
Prototypes
(3/3/50)

Requirements Specification Module
Stage 3 ~ Definition of Requirements

Figure 3

Fig. B4 Stage 3 – Definition of requirements

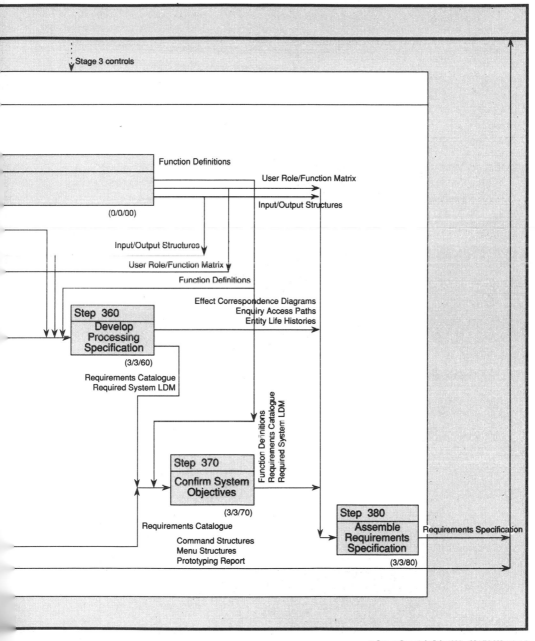

Stage 3 controls

Function Definitions

User Role/Function Matrix

Input/Output Structures

(0/0/00)

Input/Output Structures

User Role/Function Matrix

Function Definitions

Effect Correspondence Diagrams
Enquiry Access Paths
Entity Life Histories

Step 360

Develop Processing Specification

(3/3/60)

Requirements Catalogue
Required System LDM

Function Definitions
Requirements Catalogue
Required System LDM

Step 370

Confirm System Objectives

(3/3/70)

Requirements Catalogue

Command Structures
Menu Structures
Prototyping Report

Step 380

Assemble Requirements Specification

(3/3/80)

Requirements Specification

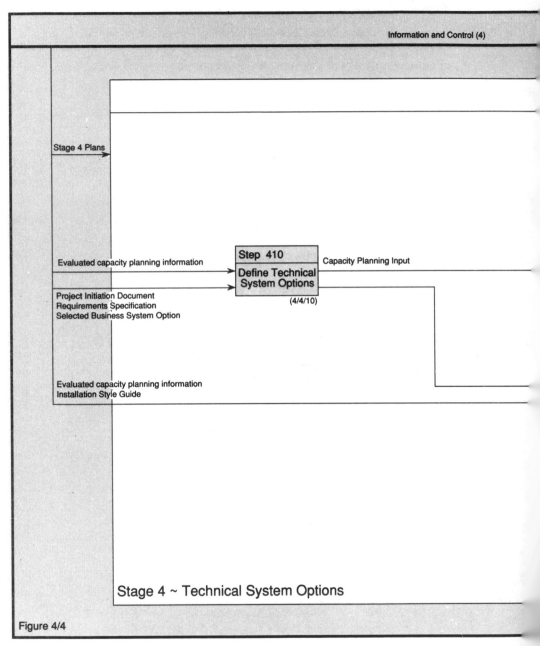

Fig. B5 Stage 4 – Technical system options

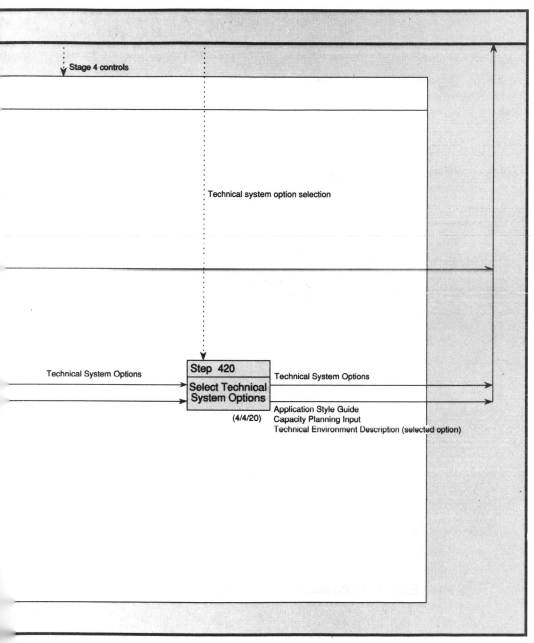

Stage 4 controls

Technical system option selection

Technical System Options

Step 420

Select Technical System Options

(4/4/20)

Technical System Options

Application Style Guide
Capacity Planning Input
Technical Environment Description (selected option)

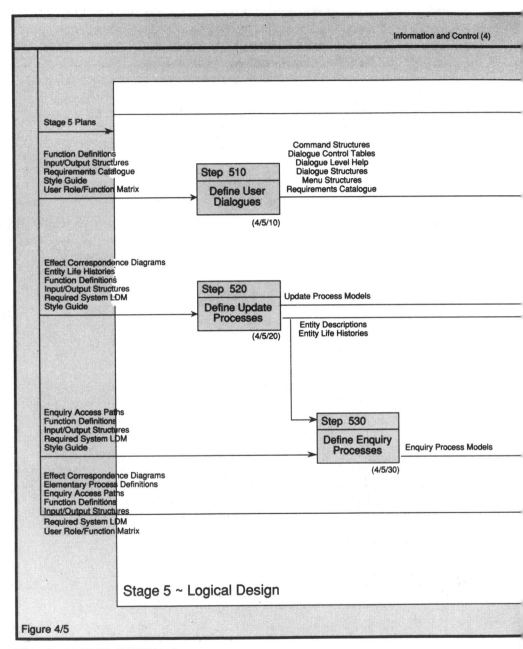

Information and Control (4)

Stage 5 Plans

Function Definitions
Input/Output Structures
Requirements Catalogue
Style Guide
User Role/Function Matrix

Command Structures
Dialogue Control Tables
Dialogue Level Help
Dialogue Structures
Menu Structures
Requirements Catalogue

Step 510

Define User Dialogues

(4/5/10)

Effect Correspondence Diagrams
Entity Life Histories
Function Definitions
Input/Output Structures
Required System LDM
Style Guide

Step 520

Define Update Processes

(4/5/20)

Update Process Models

Entity Descriptions
Entity Life Histories

Enquiry Access Paths
Function Definitions
Input/Output Structures
Required System LDM
Style Guide

Step 530

Define Enquiry Processes

(4/5/30)

Enquiry Process Models

Effect Correspondence Diagrams
Elementary Process Definitions
Enquiry Access Paths
Function Definitions
Input/Output Structures
Required System LDM
User Role/Function Matrix

Stage 5 ~ Logical Design

Figure 4/5

Fig. B6 Stage 5 – Logical design

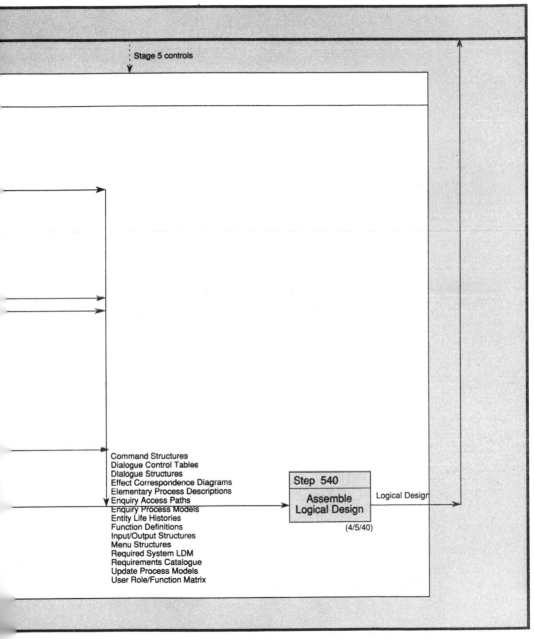

Stage 5 controls

Command Structures
Dialogue Control Tables
Dialogue Structures
Effect Correspondence Diagrams
Elementary Process Descriptions
Enquiry Access Paths
Enquiry Process Models
Entity Life Histories
Function Definitions
Input/Output Structures
Menu Structures
Required System LDM
Requirements Catalogue
Update Process Models
User Role/Function Matrix

Step 540

Assemble
Logical Design

(4/5/40)

Logical Design

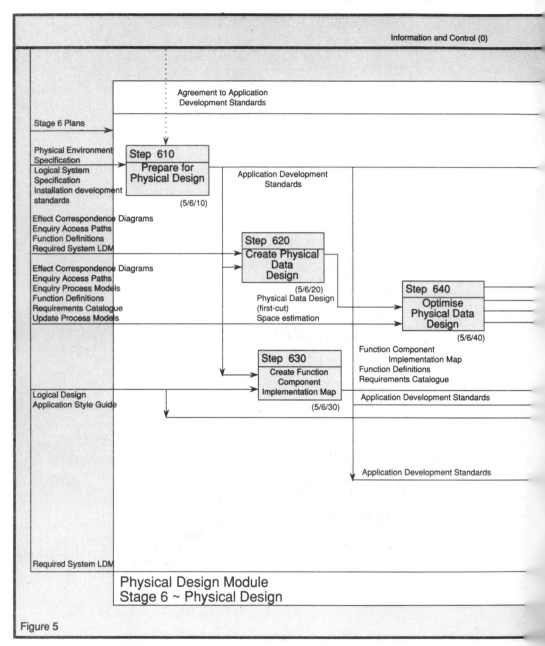

Fig. B7 Stage 6 – Physical design

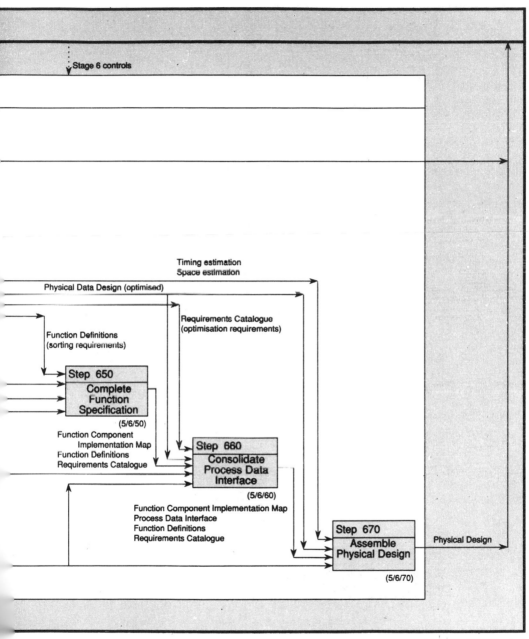

Stage 6 controls

Timing estimation
Space estimation

Physical Data Design (optimised)

Requirements Catalogue
(optimisation requirements)

Function Definitions
(sorting requirements)

Step 650
Complete Function Specification
(5/6/50)

Function Component
Implementation Map
Function Definitions
Requirements Catalogue

Step 660
Consolidate Process Data Interface
(5/6/60)

Function Component Implementation Map
Process Data Interface
Function Definitions
Requirements Catalogue

Step 670
Assemble Physical Design
(5/6/70)

Physical Design

Bibliography

Anderson, R. (1989) *Development of Business Information Systems*. Blackwell Scientific Publications, Oxford.

Ashworth, C. and Goodland, M. (1990) *SSADM: A Practical Approach*. McGraw Hill, London.

Avison, D.E. and Fitzgerald, G. (1988) *Information Systems Development: Methodologies, Techniques and Tools*, Blackwell Scientific Publications, Oxford.

Benyon, D. (1990) *Information and Data Modelling*. Blackwell Scientific Publications, Oxford.

Checkland, P (1981) *Systems Thinking, Systems Practice*. Wiley.

Clare, C. and Loucopoulos, P. (1987) *Business Information Systems*. Blackwell Scientific Publications, Oxford.

Date, C.J. (1986) *An Introduction to Database Systems* (4th Edit.). Addison Wesley, London.

De Marco, T. (1978) *Structured Analysis and Systems Specification*. Yourdon, New York.

Gane, C. and Sarson, T. (1979) *Structured Systems Analysis: Tools and Techniques*. Prentice Hall, New Jersey.

Jackson, M. (1983) *Systems Development*. Prentice Hall, New Jersey.

Longworth, G. and Nicholls, D. (1987) *SSADM Manual. Volume 1: Tasks and Terms*. NCC Blackwell, Oxford.

Longworth, G. and Nicholls, D. (1987) *SSADM Manual. Volume 2: Techniques and Documentation*. NCC Blackwell, Oxford.

Mason, D. and Willcocks, L. (1987) *Intermediate Systems Analysis*. Blackwell Scientific Publications, Oxford.

McDermid, D.C. (1990) *Software Engineering for Information Systems*. Blackwell Scientific Publications, Oxford.

Mumford, E. and Henshall, D. (1979) *A Participative Approach to Computer Systems Design*. Associated Business Press.

Myers, G. (1978) *Composite/Structured Design*. Van Nostrand Reinhold, New York.

Olle, T.W. et al (1988) *Information Systems Methodologies*. Addison Wesley, Wokingham, England.

Ratcliffe, B. (1988) *Software Engineering: Principles and Methods*. Blackwell Scientific Publications, Oxford.

Reisig, W. (1985) *Petri Nets, an Introduction*. Springer Verlag, Berlin.

Rozenberg, G. (Ed.) (Annual) *Advances in Petri Nets*. Springer Verlag, Berlin.

Yourdon, C. and Constantine, L. (1978) *Structured Design*. Yourdon, New York.

Index